D0909346

Algal Cultures and Phytoplankton Ecology

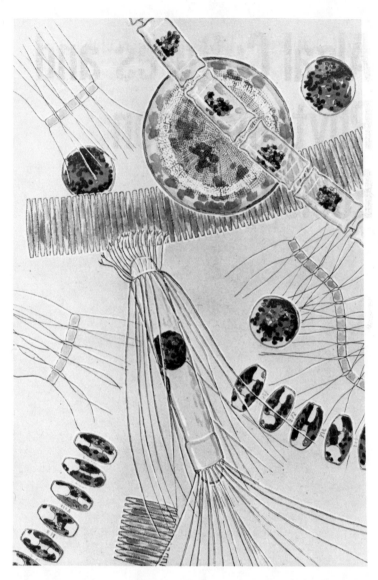

Phytoplankton diatoms from the Weddell Sea (60° 42′ S, 45° 36′ W), January 1966; *Fragilariopsis* sp. (ribbon), *Thalassiosira antarctica* (small pillbox type, single or joined in chains by mucilage strands), *Corethron criophilum* (long spiny cylinder), *Coscinodiscus stellaris* (large pillbox type), *Biddulphia striata* (chains of cells with short spines) and *Chaetoceros schimperianum* (chains of small cells with long spines). The contracted protoplasts indicate that the *Corethron* and *Biddulphia* were suffering ill effects from collection. Water-color drawing by G. E. Fogg (X 200).

Algal Cultures and Phytoplankton Ecology

SECOND EDITION

G. E. Fogg

The University of Wisconsin Press

Published 1965, 1975
The University of Wisconsin Press
Box 1379, Madison, Wisconsin 53701

The University of Wisconsin Press, Ltd.
70 Great Russell Street, London

Second Edition 1975

Printed in the United States of America

For LC CIP information see the colophon

ISBN 0-299-06760-2

To My Wife

Contents

Illustrations

Tables

Preface to the Second Edition

In the last ten years our knowledge of the behavior of phytoplankton, both in laboratory culture and in its natural habitats, has increased greatly. At the same time there has been growing realization that such knowledge is of great value if effective and ecologically acceptable ways of dealing with pollution and of using aquatic environments as sources of food are to be devised. Revision of this book has thus seemed desirable. In doing this, many sections have been completely rewritten and some increase in length has been unavoidable, but I have again aimed at brevity and clarity rather than comprehensive treatment.

My thanks are due to Dr. A. E. Walsby, for his many valuable comments on the manuscript, and to Mrs. Pat Jones and Mrs. Eileen Jackson, for their help in preparing it for the press. I am also grateful to those, to whom acknowledgment by name is made elsewhere in the book, who have so generously supplied illustrations.

G. E. Fogg

Marine Science Laboratories
(University College of North Wales)
Menai Bridge, Isle of Anglesey
March 1975

Algal Cultures and Phytoplankton Ecology

1

Introduction

The distribution and variations in abundance of the phytoplankton of lakes and seas are as yet difficult to account for except in vague and general terms. Since this microscopic floating plant life is the primary producer of the organic matter on which all other forms of life in any large body of water depend, the inadequacy of our knowledge, lamentable enough in itself, is reflected through the entire fields of freshwater and marine biology. On the economic plane this lack of information limits the efficiency with which both fresh water and the sea can be exploited as sources of food and hampers our ability to control the effects of pollution. Undoubtedly, more extensive and detailed observations of phytoplankton populations and environmental factors in natural waters will contribute greatly to our understanding of what is taking place, but studies in the laboratory with cultures of phytoplankton algae under precisely controlled conditions also have an indispensable part to play. As in ecology generally, full understanding of the growth of organisms in their natural habitats will be achieved only by the synthesis of the results of physiological and biochemical investigations and the results of field studies. The obstacles in the way of such synthesis may perhaps be less in the study of plankton than in other branches of ecology. In the present state of our ignorance, at least, both the environment and the organisms seem of less intractable complexity than those in a terrestrial community. If this is so, it should not be too difficult to relate results obtained under the artificial conditions of the laboratory to the natural situation. Equally, attempts to carry out controlled experiments *in situ* in lakes or in the sea to discover the basic factors at work do not seem altogether futile.

It is the purpose of this book to outline the principal features of the growth of algae in culture and to discuss how far they assist in the understanding of phytoplankton ecology. Both freshwater and marine phytoplankton will be considered, for it is reasonable to suppose that the factors controlling plankton growth are basically similar in fresh and salt waters, and that, *mutatis mutandis,* the principles discovered should apply in either environment. Some discussion of phytoplankton photosynthesis is inevitable, but it is not intended to give here any extensive account of the already well-documented topic of primary productivity (for reviews see Strickland, 1965, 1972; Steemann Nielsen, 1963, 1964*a*; Goldman, 1965; Vollenweider, 1974; Costlow, 1971; Platt and Subba Rao, 1973; Fogg, in the press *b*).

It will be as well to begin by emphasizing that there are many pitfalls and that we must be extremely cautious in applying results obtained with laboratory cultures to the interpretation of events in a lake or in the sea. In the first place, the algae most studied in the laboratory, predominantly species of *Chlorella,* are largely soil or non-planktonic aquatic forms. Cultures of these algae are sometimes described as planktonic, but this is merely meant to imply that the cells can be grown in free suspension under artificial conditions. The variety in physiological behavior found even in the genus *Chlorella* makes it clear that information obtained with cultures of these forms does not necessarily hold for other algae. The growth and biochemical characteristics of different strains of *Chlorella* are diverse (Winokur, 1948; Shihira and Krauss, *circa* 1965; Kessler, 1972). Some data illustrating this point are given in Figure 8, p. 25. The only truly planktonic strain which appears to have been studied in culture differs from most other strains of *Chlorella* in that it will apparently grow only photosynthetically and not on organic substrates in the dark (Fogg and Belcher, 1961). Algae belonging to other classes may be expected to be even more different, and the fact that comparatively few successes have thus far been obtained in growing species of planktonic algae in culture emphasizes the point that their growth requirements are not necessarily the same as those of the commonly studied kinds.

Natural waters themselves are unsatisfactory for sustained growth of algae in the laboratory, mainly because some essential nutrients are usually present in only trace amounts, their concentrations depending on dynamic equilibria which are disturbed as soon as the water is collected. Natural waters supplemented with various nutrients have been much used when the object has been only to produce algal material, precise knowledge of the conditions affecting its growth being unnecessary. The best known of such media is "Erd-Schreiber" solution, which is natural sea water supplemented with soil extract, nitrate, and phosphate. Artificial media for freshwater algae have been developed empirically, the simple solutions of a few mineral salts used by pioneers such as Benecke and Beijerinck being modified by variation in the proportions of the major nutrients and addition of trace elements, as these were discovered to be essential for healthy growth. These phases in the development of media have been summarized by Pringsheim (1946) in his book, *Pure Cultures of Algae*. The mineral requirements of algae have been reviewed by O'Kelley (1968), and consideration of the rôles of individual elements will be found in the books edited by Lewin (1962) and Stewart (1974). Recipes for media are given by Stein (1973).

Usually investigators have been content, when selecting a medium, to test a number of well-established recipes and use the most suitable with a minimum of modification. Only a few workers (*e.g.*, Rodhe, 1948; Krauss, 1953; Miller and Fogg, 1957) have undertaken determinations of growth in series of media in which the concentrations and proportions of the constituents were systematically varied. Generally, the criterion of the usefulness of a medium has been the final yield of algal material which the medium gives, and consequently the inorganic compositions of the most popular media bear little resemblance to those of natural waters. Chu (1942) was a pioneer among those who set out to devise media having some resemblance to those in which algae grow naturally. His highly successful medium No. 10 is comparable in composition and degree of dilution to the water of a eutrophic lake (Table 1). More recently there have been great improvements in artificial media for marine phytoplankton. These media bear a general but not necessarily close resemblance to sea water, in terms of concentrations of major ions. They also contain chelating agents to maintain adequate amounts of trace elements in solution, and organic growth factors. Some chemical considerations in the design of synthetic media have been given by Droop (1961*b*), and recipes have been given by Provasoli *et al.* (1957, see Table 1), Provasoli (1968), Droop (1969), and Stein (1973). Requirements for organic growth factors are now known to be common among algae. Thiamine (vitamin B_1), cobalamin (vitamin B_{12}), and biotin are of most general importance, but some species require particular amino acids (Droop, 1962b; Provasoli, 1963, 1971). A rough estimate is that 70 per cent of planktonic algae require vitamin B_{12}. Most studies on vitamin requirements have been carried out with marine forms but Moss (1973*b*) has shown that vitamin B_{12} is essential for many freshwater phytoplankton species although none of those which he studied required thiamine or biotin. Droop (1961*b*, 1962*a*) has paid particular attention to the problems of chelation of trace elements, *p*H buffering, and the poising of oxidation-reduction potential, some or all of which physicochemical characteristics of a medium may be more crucial for the growth of a phytoplankton species than are the proportions of the major ions. The compositions of three synthetic media are given in Table 1, together with typical compositions of fresh water and sea water. Artificial media, simulating natural waters but of precisely known composition, have reached a high level of sophistication. Nevertheless, we cannot assume exact correspondence. Natural waters cannot be looked on as simple solutions of mineral salts plus certain definite organic substances having chelating or growth-promoting properties. They normally contain a relatively high concentration of dissolved organic matter, an average value being about 5 mg/liter, and we know extremely little about the nature of the substances which make up this total (Duursma, 1965; Hood, 1970). Undoubtedly, in chelation and in supplying organic growth factors, this dissolved organic material plays rôles which can be performed equally well by the chemically defined substances in artificial media, but it may also have other biological effects of which we are at present ignorant. An often-quoted example which indicates the importance of such unknowns is the finding of Rodhe (1948; see also Mackereth, 1953) that in artificial culture media concentrations of at least 0.040 mg/liter of phosphorus were required to sustain maximum growth rates of *Asterionella formosa*, whereas in natural lake water as little as 0.002 mg/liter sufficed. Evidently there is some factor present in lake water which enables *A. formosa* to make use of extremely low concentrations of phosphate, but what the nature of this factor is remains unknown.

We are also unable at present to account for the difference between the "good" and "bad" sea waters studied by Wilson (Wilson and Armstrong, 1958). Johnson (1963b) found that such difference persists even after supplementation with phosphate, nitrate, silicate, trace metals, chelating materials, vitamin B_{12}, and thiamine. He concluded that one or more unknown labile factors which either promote or inhibit growth are important in sea water.

It must also be recognized that enclosing a culture in a container itself introduces differences which may be significant. The surface presented by the vessel, relative to the volume of medium, is vastly in excess of that presented by the normal amount of particulate matter in a natural water. No surface can be totally inert but must modify conditions to some extent by adsorption of solutes from the medium, if not by release of traces of soluble matter as well. It is well established that development of bacteria in water samples is proportional to the surface area presented by the containing vessel (ZoBell, 1946), so that we may expect the relationship between these microorganisms and the phytoplankton to be altered in culture. Materials such as glass may not be transparent to wavelengths of ultraviolet light which can penetrate to some extent into water. Ilmavirta and Hakala (1972) found striking differences between the photosynthetic behavior of similar algal suspensions held in Jena glass and acrylic plastic bottles which might be ascribed to differences in transparency. Furthermore, the small-scale turbulence patterns, which are so important for the supply of nutrients to the cells (see p. 71), are likely to be rather different in a culture vessel from those in open water. That these patterns can be critical for the survival of all but the most robust of algae is illustrated by the findings of Fogg and Than-Tun (1960) that the growth of *Anabaena cylindrica* was doubled when culture flasks were shaken at 90 instead of 65 oscillations per minute and entirely prevented at 140 oscillations per minute. A culture flask designed by Walsby (1967), which when shaken imparts a regular swirling motion to the medium, has proved particularly suitable for the culture of such algae.

The initial objective of the laboratory worker is usually to isolate a phytoplankton organism in pure (or *axenic*) culture. This is commonly a difficult task and is perhaps achieved far less often than is claimed, since it is impossible to prove a negative; tests for the presence of bacteria, even if rigorous, may fail to demonstrate kinds with unusual

TABLE 1

The compositions of a typical eutrophic freshwater, seawater, and three synthetic culture media for algae (amounts expressed in mg/liter)

	Kettle Mere, Shropshire, England (Gorham, 1957)	Medium No. 10 of Chu (1942) modified by addition of iron as citrate	*Monodus* standard medium (Miller and Fogg, 1957)	Sea water (Golderg, in Hill, 1963; Sverdrup *et al.*, 1942)	Synthetic seawater medium ASP 2 (Provasoli *et al.*, 1957)
Na	7.6	18.1	460	10,500	7,050
K	8.6	4.5	90	380	313
Ca	23.2	9.7	25	400	100
Mg	2.9	2.5	19	1,350	440
HCO_3	34.8	23.0	–	140*	–
Cl	13.9	–	45	19,000	10,400
SO_4	26.8	9.7	77	2,660	1,930
$NO_3–N$	0.05	6.8	280	0.001–0.60	8.2
$PO_4–P$	0.004	1.8	37	0.07	0.9
SiO_2	1.0	12.3	–	6.4	3.2
Fe	–	0.18†	1.1†	0.01	0.8

B	—	—	0.2	4.6	6.0
Mn	—	—	0.2	0.002	1.2
Mo	—	—	0.2	0.01	–
Co	—	—	0.02	0.0005	0.003
Cu	—	—	0.02	0.003	0.0012
Zn	—	—	0.02	0.01	0.15
tris (hydroxymethyl) aminomethane	—	—	–	–	1,000
sodium ethylenediamine tetraacetate	—	—	–	–	30
vitamin B_{12}	—	—	–	–	0.002
thiamine hydrochloride	—	—	–	–	0.5
nicotinic acid	—	—	–	–	0.1
calcium pantothenate	—	—	–	–	0.1
p-aminobenzoic acid	—	—	–	–	0.010
biotin	—	—	–	–	0.001
inositol	—	—	–	–	5
folic acid	—	—	–	–	0.002
thymine	—	—	–	–	3

* at pH 7.0
† as ferric citrate

nutritional requirements. Clearly, it is necessary to eliminate other microorganisms from cultures if one is to study the nutrition and metabolism of a particular species, but then the problem arises as to how far information thus gained is applicable to natural conditions in which the species is associated and interacting with others. It is apparent that there are close interrelations between plankton algae and their associated bacteria. Thus, *Asterionella japonica* was found to grow satisfactorily in culture as long as bacteria were present but ceased to grow when these were eliminated, even though a variety of possible organic growth factors was provided (Kain and Fogg, 1958). Johnston (1963*b*) found that whereas bacteria-free *Skeletonema costatum* grew poorly in various samples of sea water enriched with nitrate, phosphate, silicate, and chelated trace metals, it grew distinctly better in the same media if bacteria were present. This effect seemed largely due to production of vitamin B_{12} by the bacteria. It seems that, having studied pure cultures, physiologists and biochemists must next investigate cultures containing more than one species. Such cultures, however, are difficult to achieve. Although two phytoplankton species may apparently coexist in equilibrium in a natural water, almost invariably one will completely outgrow the other if brought into culture.

Finally it should be pointed out that the physical environment in which cultures are maintained in the laboratory does not necessarily correspond to that in which phytoplankton grows naturally. Temperature conditions need not be widely different, since temperatures in large bodies of water are comparatively stable. Laboratory cultures are customarily incubated at constant temperature although the temperature is often considerably higher than that which algae normally encounter in their original habitat, since constant temperatures above room temperature are easier to achieve. Underwater illumination, on the other hand, is extremely difficult to imitate. Gradients in intensity of total radiation çan be paralleled fairly well by use of suitable neutral filters, but light quality is difficult to match. The spectral composition of light penetrating the water changes with depth, is different for different kinds of water, and varies according to the weather (Jerlov, 1966, 1968). Except at the immediate surface it is very different from ordinary sunlight, and it is difficult to duplicate even approximately with artificial sources (Jitts *et al.,* 1964; Eppley and Strickland, 1968). Since the quantum efficiency of photosynthesis varies with wavelength, and since the action spectra for various photochemical effects are different, these discrepancies cannot be ignored (Dring, 1971). Evidence that far-red light has effects on the growth rates of phytoplankton species has been reported by Lipps (1973). In addition, natural light fields have a diurnal rhythm of variation both in quality and in quantity. Laboratory cultures are commonly incubated in continuous illumination. The inadequacies of this practice for forms such as *Hydrodictyon,* which, as Pirson (1957) has shown, requires regular alternation of light and dark if it is to survive, are obvious, but it is likely that many subtle modifications of growth and metabolism induced by continuous illumination have as yet escaped notice. Alternation of periods of light of fixed quality and intensity with periods of complete darkness may be an approximation to natural conditions sufficient for many purposes, but incubation of cultures *in situ* in the sea or lakes, inconvenient though it is, is perhaps a method that ought to be used more frequently.

The Characteristics of Algal Growth in Cultures of Limited Volume

General accounts of the laboratory culture of algae have been given by Myers (1962a), Droop (1969), and Stein (1973). The most usual kind of culture in experimental work is one in which a limited volume of medium containing the necessary inorganic and organic nutrients is inoculated with a relatively small number of cells and then exposed to suitable conditions of light, temperature, and aeration. Increase in cell numbers in such a batch culture follows a characteristic pattern, represented in Figure 1, in which the following phases may usually be recognized: (1) a lag or induction phase, in which no increase in cell numbers is apparent; (2) an exponential phase, in which cell multiplication is rapid and numbers increase in geometric progression; (3) a phase of declining relative growth; (4) a phase in which cell numbers remain more or less stationary; and (5) a death phase. Sometimes one or more of these phases may be so curtailed as to be scarcely recognizable.

A lag in cell multiplication may be apparent rather than real if a large proportion of the cells inoculated is not viable. Cell numbers will then remain nearly stationary until the progeny of the cells capable of dividing reach a number comparable with the total inoculated (Fig. 2). This has been shown to be the most important cause of the lag when *Monodus subterraneus* is subcultured after a period of dark incubation

12

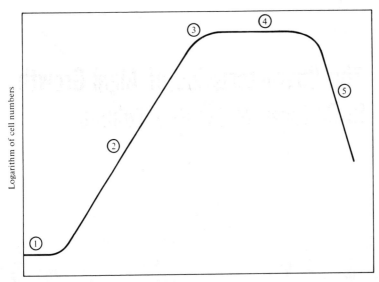

Figure 1. The characteristic pattern of growth shown by a unicellular alga in a culture of limited volume; (1) lag phase, (2) exponential phase, (3) phase of declining relative growth rate, (4) stationary phase, (5) death phase.

(Belcher and Miller, 1960). On the other hand, a majority of the cells inoculated may be viable but not in a condition to divide immediately. Especially if the parent culture was an old one, enzymes may have been inactivated, and concentrations of metabolites may have decreased to a level insufficient for cell division, so that a period of reconstitution is necessary before active growth can begin. This gives rise to a true lag phase, which, of course, may be superimposed on an apparent lag due to non-viable cells. It may be noted that synthesis of cell material does not necessarily show a lag and that cells may increase in size during the lag phase.

Much of Hinshelwood's (1946) discussion of the kinetics of bacterial lag phase is of general application to other organisms and the reader is referred to his book for mathematical formulation of the relations described below for algae. The most detailed study of the lag phase in algae appears to be that on *Phaeodactylum tricornutum* (*Nitzschia closterium* forma *minutissima*) by Spencer (1954). With this species, as with *Anabaena cylindrica* (Fogg, 1944) and probably with all algae, the

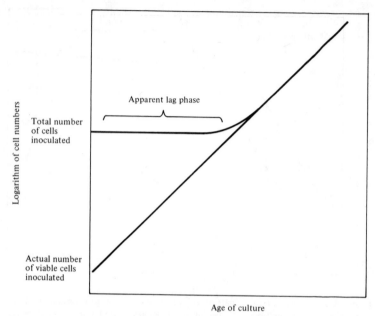

Figure 2. Production of an apparent lag phase as a result of a high proportion of the cells inoculated being non-viable.

length of the lag was dependent on the age of the inoculum, diminishing as this entered the exponential phase of growth, being zero if the inoculum had been growing exponentially, then increasing according to the duration of the stationary phase (Fig. 3). This fits in with the picture of the lag phase as a period of restoration of enzyme and substrate concentrations to the levels necessary for rapid growth. It is to be noticed that cells subcultured during the early part of the exponential phase showed a short lag. If small inocula were used, even cells taken when exponential growth was well established showed a lag in fresh medium. A presumably related finding is that cultures of certain planktonic blue-green algae can be established only from large inocula (Gerloff *et al.*, 1950). Eberley (1967) found that decreasing amounts of inoculum caused the lag phase in cultures of *Oscillatoria agardhii* to increase from zero to over 10 days. Taken in conjunction with the fact that under some conditions cells inoculated into re-enriched medium from an old culture showed less lag than similar cells

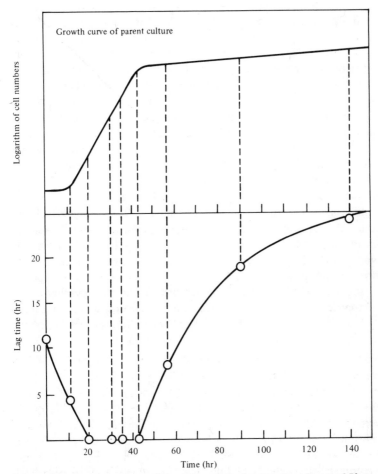

Figure 3. Variation with age of inoculum of the lag time in cultures of *Phaeo-dactylum tricornutum* grown at 19° C with continuous illumination and aeration. After C. P. Spencer, Studies on the Culture of a Marine Diatom, J. *mar. biol. Ass. U.K.* (1954), *33*:285, fig. 14 (Cambridge University Press).

in fresh medium, these results suggest that some diffusible factor produced by the cells themselves is necessary for optimum growth. An alternative explanation is that some toxic factor present in the medium is inactivated by the metabolic activity of the cells. It is well established that the former state of affairs occurs with certain bacteria. Both *Bacterium lactis-aerogenes* and pneumococci require a certain minimum concentration of carbon dioxide in the medium for growth. In fresh medium there is a lag until this concentration is built up by respiration, but if air free of carbon dioxide is blown through the culture, the lag is prolonged indefinitely (for references see Hinshelwood, 1946).

It is possible that for some plankton algae glycolic acid, $CH_2OH \cdot COOH$, is a diffusible factor that must be built up to a certain concentration in the medium before growth can begin. Tolbert and Zill (1956) found that in short-term photosynthesis experiments using ^{14}C-labeled bicarbonate, relatively considerable amounts of labeled glycolic acid were liberated by *Chlorella*. This observation has been amply confirmed, and it appears that glycolic acid is derived from ribulose diphosphate, the carbon dioxide acceptor in the photosynthetic fixation cycle, production of glycolic acid being increased at low carbon dioxide concentrations, when inhibitors of its further metabolism are present, or at high oxygen concentrations (Tolbert, 1974; Fogg, in the press *a*). Not all algae excrete glycolate in appreciable amounts nor do all species take it up when it is supplied in the external medium. Tolbert (1974) has shown, by studies with ^{14}C-labeled glycolate, that it is assimilated to a slight extent only, or not at all, by some species, particularly those known to excrete it. However, other species are capable of utilizing glycolate as a carbon source and a few of these also excrete it. With species which both excrete and utilize the substance there may be a quasi-equilibrium between the intra- and extracellular concentrations of glycolate. This suggests that when photosynthesis begins in fresh medium there must first of all be a lag while an equilibrium concentration of glycolic acid is established in the medium, and that only when this is achieved can the products of carbon fixation become available for growth. This idea is supported by the finding that the lag shown by small inocula of a planktonic strain of C. *pyrenoidosa* at limiting light intensities is abolished by the addition to the medium of concentrations of glycolic acid of the order of 1 mg/liter. Equivalent additions of glucose or of other organic acids do not have a similar effect. At saturating light intensities or with heavy inocula the lag phase of this strain of *Chlorella* is much reduced, as under these conditions the cells are able to establish the necessary concentration of extracellular glycolic acid rapidly (Sen and Fogg, 1966; see Fig. 4). Similar reduction of the lag phase by low concentrations of glycolic acid has been found for the marine diatoms *Ditylum brightwellii* (Paredes, 1967/68a) and *Nitzschia closterium* (Tokuda, 1966) and the flagellate *Micromonas squamata* (Paredes, 1967/68b). The lag phase of the diatom *Skeletonema costatum,* which is known to liberate glycolate as an extracellular product, is prolonged indefinitely in the presence of alumina, a powerful adsorbent of glycolate (Fig. 5).

A lag phase may also be exhibited when algal cells are subcultured into medium containing high concentrations of some particular substance. Spencer (1954) found that *Phaeodactylum* cells from old phosphate-deficient cultures subcultured into fresh medium with increased phosphate concentration showed a much increased lag phase as compared with that of similar cells introduced into medium with low phosphate concentration. Prolonged lag phases have been observed when *Anacystis nidulans* is inoculated into media containing sublethal concentrations of antibiotics (Fig. 6). Such lag phases may represent the period needed for the reconstitution of enzymic constituents of the cells to meet the changed metabolic circumstances (Hinshelwood, 1946), the period needed for degradation of the antibiotic, or, as is probably the case for the adaptation of *Anacystis* to antibiotics, the period needed for the multiplication of mutant forms (Kumar, 1964).

During the exponential phase, growth results in the production of more material, which is itself capable of growth, so that the actual rate of growth accelerates continuously. This can be represented by the expression

$$W = W_o e^{kt},$$

in which W_o is the total amount of cell material in the culture at zero time, W the amount after a period of time, t, e the base of natural logarithms, and k (sometimes denoted by R or μ) the relative growth constant, which is a measure of the efficiency of growth. In large populations in which cell divisions are not synchronized, increase in cell numbers (N) per unit volume takes place smoothly and may be represented by the corresponding expression:

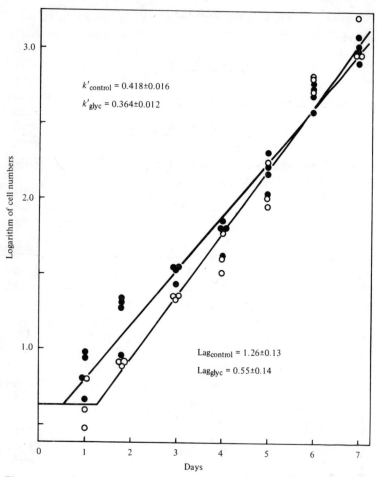

Figure 4. Growth of a planktonic strain of *Chlorella pyrenoidosa* from a small inoculum at limiting light intensity in the presence (•) and absence (o) of 1 mg per liter of glycolic acid. From N. Sen and G. E. Fogg, Effects of glycolate on the growth of a planktonic *Chlorella, J. exp. Bot.* (1966), *17*:417–425, fig. 1 (Clarendon Press, Oxford)

$$N = N_o e^{kt}$$

whence

$$k = \frac{\log_e N - \log_e N_o}{t} \, ,$$

or, if logarithms to the base 10 are preferred,

$$k' = \frac{\log_{10} N - \log_{10} N_o}{t} \, .$$

From this, the mean doubling time G (which equals the mean generation time if the cells divide into two) will be

$$G = \frac{0.301}{k'} \, .$$

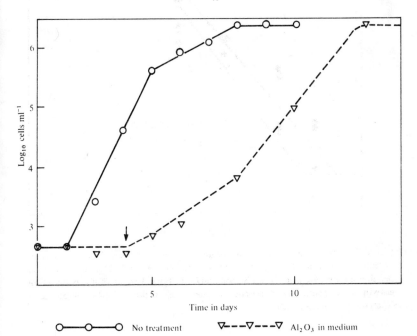

Figure 5.   Growth of the marine diatom *Skeletonema costatum* at 6000 lux, with ($\triangle$) and without ($\circ$) 2 g alumina in 50 ml of medium. The arrow marks the point at which the culture was decanted off from the alumina. Data of Pant, 1973.

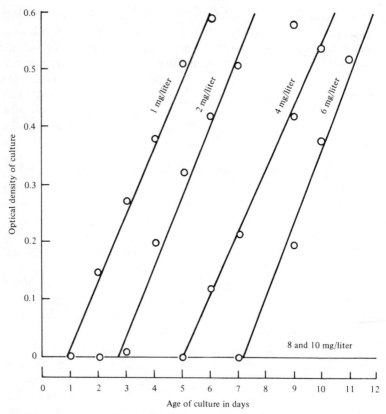

Figure 6. Growth of *Anacystis nidulans* in the presence of different concentrations of chloramphenicol. Data of Kumar, 1964.

Many authors prefer to use logarithms to the base 2, in which case the reciprocal of the relative growth rate is equal to the doubling time. Values of k based on different measures of growth, such as cell numbers, volume of algal material, or cell nitrogen, must clearly be approximately the same but need not be identical, since the mean generation time may remain constant while the mean cell volume or dry weight diminishes.

Values of relative growth rates have been collected together by Hoogenhout and Amesz (1965) and further references are given by Eppley and Strickland (1968). A selection for various planktonic and non-planktonic algae is given in Table 2. Measured under standard conditions, k' is usually rather constant for a particular species. Its value may depend on a variety of physiological and metabolic factors, but size is generally of most importance, since it determines the surface/volume ratio and thus the relative rate at which materials for growth can be taken into the cell. It is a general observation that small species grow faster than large ones. For example, in Table 2 it will be seen that the small forms, *Chlorella* and *Anacystis,* grow much more rapidly than the large ones, *Ceratium tripos* and *Coscinodiscus* sp. The very low relative growth rate of *Gloeotrichia echinulata* is attributable to the large size of its colonies, which may be up to 7 mm in diameter. For a single species a good correlation may be found between relative growth rate and surface/volume ratio although the relation is not a linear one, the expected relationship if no other factor were concerned (Fig. 7). However, in comparing different species the correlation is further obscured since they cannot always be grown under identical conditions and, as Paasche (1960*a*) has pointed out, the correlation may not hold for cells with large vacuoles. In such cells, that part of the cytoplasm in which photosynthesis occurs may have just as much surface available for exchange of materials as would the same volume of cytoplasm in a smaller cell lacking vacuoles.

For a given species the relative growth constant is a function of temperature, light intensity, and other environmental factors. As Table 2 and Figure 8 show, it increases with temperature. The Q_{10} of k is usually from 2 to 4 until unfavorably high temperatures are reached. Optimum temperatures are generally between 20° and 25° C, but the thermophilic strains of *Chlorella* and *Anacystis nidulans* grow best at about 40° C. However, optimum temperatures may vary with light intensity and concentration of certain nutrients, and adaptation to higher or lower temperatures may sometimes occur. Hutner *et al.* (1957) succeeded in growing *Ochromonas malhamensis* above 35° C—the maximum temperature tolerated in the normally adequate basal medium—by supplying thiamine and vitamin B_{12} at 300 times the normal concentrations. A similar interaction between nutrient concentration and temperature has been reported by Maddux and Jones (1964). They found that the optimum temperatures for growth of *Nitzschia closterium* and *Tetraselmis* sp. were lower when a medium with nitrate and phosphate concentrations similar to those found in

TABLE 2

Relative growth constants, k', in $\log_{10}$ day units, and mean doubling times, G, in hours, of various planktonic and non-planktonic algae grown in continuous light of intensities approximately saturating for photosynthesis

Species	k'	G	Temp. °C	Reference
Chlorophyceae				
Chlorella pyrenoidosa	0.93	7.75	25	Sorokin, 1959
Nannochloris sp.	1.35	5.3	33	Thomas, 1966
Dunaliella tertiolecta	0.30	24	16	McLachlan, 1960
Scenedesmus quadricauda	1.23	5.9	25	Österlind, 1949
Xanthophyceae				
Monodus subterraneus	0.074	97.7	15	Fogg et al., 1959
	0.191	37.8	20	Fogg et al., 1959
	0.297	24.3	25	Fogg et al., 1959
	0.169	42.7	30	Fogg et al., 1959
Chrysophyceae				
Monochrysis lutheri	0.48	15.3	20—25	Antia and Kalmakoff, 1965

Haptophyceae				
Isochrysis galbana	0.24	30.2	20	Kain and Fogg, 1960
Cricosphaera (Syracosphaera) carterae	0.36	20.1	18	Parsons et al., 1961
Bacillariophyceae				
Asterionella formosa	0.75	9.6	20	Lund, 1949
Asterionella japonica	0.52	13.9	20–25	Kain and Fogg, 1960
Phaeodactylum tricornutum	0.72	10.0	25	Spencer, 1954
Skeletonema costatum	0.55	13.1	18	Parsons et al., 1961
Coscinodiscus sp.	0.20	30.0	18	Parsons et al., 1961
Coscinodiscus pavillardii	0.35	20.9	28	Findlay, 1972
Chaetoceros sp.	1.81	4.0	29	Thomas, 1966
Dinophyceae				
Amphidinium carteri	0.82	8.8	18	Parsons et al., 1961
Prorocentrum micans	0.13	55.5	20	Kain and Fogg, 1960
Ceratium tripos	0.087	82.8	20	Nordli, 1957
Cyanophyceae				
Anabaena cylindrica	0.68	10.6	25	Fogg, 1949
Anacystis nidulans	3.55	2.0	41	Kratz and Myers, 1955
Gloeotrichia echinulata	0.06	120.0	26–27	Zehnder, 1963

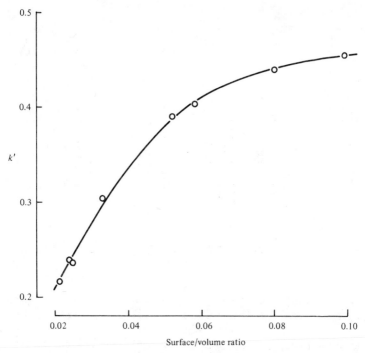

Figure 7.   The relation of relative growth constant (k') to surface/volume ratio in the tropical marine diatom *Coscinodiscus pavillardii*. Data of Findlay, 1972.

natural waters was used than they were when a medium having higher concentrations of these substances was employed.

As would be expected, the relative growth rate bears the same general relationship to light intensity as does the rate of photosynthesis, increasing proportionately to intensity when intensity is the limiting factor and being independent of intensity when saturating values are reached. However, in the relation of growth to light intensity, there is the complication that adaptation to different intensities occurs rather quickly while exponential growth is taking place. This is well illustrated by some results obtained with *Chlorella vulgaris* by Steemann Nielsen *et al.* (1962). The alga was grown under high (30 kilolux) and low (3 kilolux) light intensity in dilute culture so that there was no appreciable shading of cells by others. The curves relating rate of photosynthesis to light intensity for the two cultures are quite different (Fig. 9), the cells

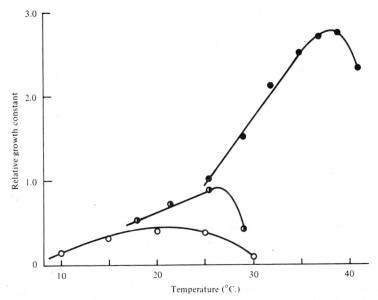

Figure 8. Relative growth constant (k') at about 15,000 lux for three strains of *Chlorella* as a function of temperature. o, *C. pyrenoidosa,* cold-water strain isolated from Torneträsk, Sweden (data of Fogg and Belcher, 1961); ⊙, *C. pyrenoidosa,* Emerson strain; •, *C. sorokiniana,* high-temperature strain (data of Sorokin and Myers, 1953). From G. E. Fogg, Survival of algae under adverse conditions, *Symp. Soc. exp. Biol.* (1969), *23*:123–142, fig. 1 (Cambridge University Press).

grown in low light being more efficient at low intensities but becoming saturated at a lower level than the cells grown in high light. The greater efficiency of the low light cells results from their higher concentration of chlorophyll, since, if photosynthesis per unit amount of chlorophyll is plotted, there is no difference between the two types when light is limiting (Fig. 10). This behavior of *Chlorella* thus corresponds to the production of "shade" and "sun" leaves by higher plants. Adaptation by the alga from one condition to the other is comparatively rapid, requiring only one division cycle, *i.e.,* about 17 hours under the experimental conditions used. The existence of such adaptation explains the discrepancy, pointed out by Myers (1951), that *Chlorella* appears to be able to produce material in photosynthesis as much as twice as fast as it can utilize it in growth. This finding was based on the

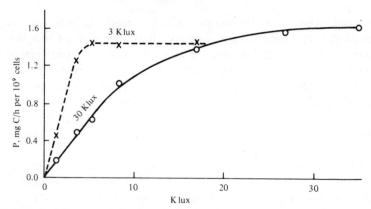

Figure 9. The rate of photosynthesis per unit number of cells as a function of light intensity for *Chlorella vulgaris* grown at 3 or 30 kilolux. After E. Steemann Nielsen, V. K. Hansen, and E. G. Jørgensen, The adaptation to different light intensities in *Chlorella vulgaris* and the time dependence on transfer to a new light intensity, *Physiol. Pl.* (1962), *15*:508, fig. 2, part *a* (Munksgaard A.S.).

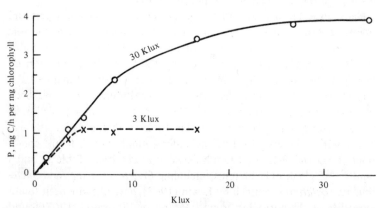

Figure 10. The rate of photosynthesis per mg chlorophyll as a function of light intensity for *Chlorella vulgaris* grown at 3 or 30 kilolux. After E. Steemann Nielsen, V. K. Hansen, and E. G. Jørgensen, The adaptation to different light intensities in *Chlorella vulgaris* and the time dependence on transfer to a new light intensity, *Physiol. Pl.* (1962), *15*:508, fig. 2, part *b* (Munksgaard A.S.).

comparison of curves showing the relation of relative growth rate and photosynthetic rate to light intensity. It is now realized that this comparison is not valid because the algal cells used in the two series of determinations were differently adapted. Those used in the growth determinations were grown at, and therefore adapted to, each intensity at which measurements were made, whereas those used in the photosynthesis experiment were all grown at the same intensity and were therefore not adapted to the different intensities at which rates of photosynthesis were determined.

Further experiments by Jørgensen (1969) have shown that although some algae, such as various species of Chlorophyceae, the xanthophycean *Monodus,* and the blue-green *Synechococcus elongatus,* resemble *Chlorella* in respect of light adaptation, others show different behavior. In this second group, which includes *Scenedesmus quadricauda* amongst the Chlorophyceae and the diatoms *Skeletonema costatum, Cyclotella meneghiniana, Nitzschia palea,* and *N. closterium,* the amount of chlorophyll per cell remains unchanged by growth at different light intensities. The only observed adaptation was that light-saturated rates of photosynthesis were higher and were attained at higher light intensities in cells grown under stronger illumination, presumably because they contained higher concentrations of the enzymes involved in the dark reactions of photosynthesis.

Many algae are able to utilize organic substrates, such as sugars and organic acids, to maintain growth in complete darkness or as the sole source or a supplementary source of carbon in the light. The algae able to grow in darkness (see the review by Danforth, 1962) are derived mainly from soil, heavily contaminated water, or littoral marine habitats (Lewin, 1963). Thus far few truly planktonic forms have been reported as being capable of this. The planktonic strain of *Chlorella pyrenoidosa* has not been found able to grow in the dark, although tested with a variety of substrates under a number of different conditions (Fogg and Belcher, 1961; Nalewajko *et al.,* 1963; Table 3), and a number of planktonic diatoms, including *Chaetoceros* spp. and *Skeletonema costatum* examined by Lewin (1963), have likewise been found incapable of chemotrophic growth. Pintner and Provasoli (1963) found that, although substrates such as lactate greatly stimulated the growth of the chrysomonad *Hymenomonas* sp., they did not support growth in the dark. Nevertheless, two species of freshwater cryptomonad were found by Wright (1964) to grow in the dark in the presence of acetate and, as we shall see later (p. 88), the possibility that other plankton algae are capable of chemotrophic growth must be considered. With several species, uptake of organic carbon at rates which are sometimes of the same order as those of assimilation of inorganic carbon in photosynthesis, have been demonstrated using radiocarbon as a tracer. This has been done for glycolic acid with *Chlorella pyrenoidosa* (Nalewajko *et al.*, 1963) and with *Skeletonema costatum* (Pant, 1973). Active uptake of several amino acids by the saltwater pond diatom *Melosira nummuloides* was demonstrated by Hellebust (1970). Two dinoflagellate species, *Gymnodinium nelsonii* and *Peridinium trochoideum*, took up glucose as well as the amino acids lysine and alanine but *Coccolithus huxleyi*, *Isochrysis galbana*, *Dunaliella tertiolecta*, *Pyramimonas* sp., and *Skeletonema costatum* did not take up any of the substrates offered at measurable rates. Hellebust (1971) found that *Cyclotella cryptica*, a littoral diatom which can grow in the dark with glucose as the sole carbon source, develops a glucose transport system in the dark but this is rapidly inactivated when the cells are transferred to the light.

In the light an organic substrate may enable an alga to attain a higher relative growth rate than would be possible without the substrate. Roach (1928), for example, found that *Scenedesmus costulatus* var. *chlorelloides* at saturating light intensities attained a maximum value for k which could not be increased by supplying glucose. At limiting light intensities, however, k was increased by a supply of glucose but never to a value more than that attained at saturating light intensity. Growth of the planktonic *C. pyrenoidosa* is similarly enhanced at limiting light intensities by glucose (Fogg and Belcher, 1961) and by glycolate (Table 3). Provision of organic substrates may increase growth

TABLE 3

Effect of glycolic acid supplied at a concentration of 1 mg per liter on the relative growth constant (k') of a planktonic strain of *Chlorella pyrenoidosa* (data of Sen and Fogg, 1966)

Light intensity	k' without glycolate	k' with glycolate
Dark	0.000	0.000
500 lux	0.110 ± 0.0150	0.208 ± 0.0085
1600 lux	0.263 ± 0.0278	0.251 ± 0.0163

at saturating light intensities by acting as an alternative source of carbon if the concentration of carbon dioxide is not saturating. However, the growth of *Ochromonas malhamensis,* a flagellate which is capable of phagotrophic as well as chemotrophic and phototrophic nutrition, is increased by organic substrates even when light and carbon dioxide are saturating for photosynthesis (Myers and Graham, 1956). Its mean generation time, with or without light, in the presence of glucose or sucrose is about 14 hours, whereas light and carbon dioxide alone will support only marginal growth with a mean generation time of about 3 days. Myers and Graham concluded that *Ochromonas* "is a very primitive animal which has retained only enough of its photosynthetic apparatus to sustain it between bites." The uptake of organic carbon by algae in the light is by direct photoassimilation and does not depend on assimilation of carbon dioxide produced by oxidation of the substrate (Wiessner, 1970). Reducing power and high energy phosphate produced by the photochemical reactions is used to convert the organic substrate into cell components. It may be noted that Pringsheim and Wiessner (1961) found this type of nutrition to be obligatory in *Chlamydobotrys* sp., which is able to grow only in the light with acetate, but not with carbon dioxide, as the carbon source.

The assimilatory mechanisms of algae appear to be saturated by extremely low concentrations of mineral ions, so that it is technically difficult to study the effects of mineral nutrient limitation on exponential growth. The effect of supplying a low concentration of a particular nutrient in a culture of limited volume is to shorten the duration of this exponential phase rather than reduce the relative growth rate. The expected relationship of relative growth constant, k, to concentration of limiting nutrient, c, is given by the following expression (Monod, 1942; Hinshelwood, 1946):

$$k = \frac{k_\infty c}{c + K_s}$$

in which K_s is a constant having the dimensions of a concentration (being numerically equal to the concentration of the nutrient giving half the maximum growth rate, k_∞). A nutrient is limiting only when c is not large compared with K_s. It follows from this relation that so long as cell numbers are sufficiently low as not to alter appreciably the concentration of the nutrient, k remains constant during growth even though the concentration is a limiting one.

Thomas and Dodson (1968) obtained values of around 0.12 μ moles per liter of phosphate for K_s with batch cultures of the tropical marine diatom *Chaetoceros gracilis* at 27° C. Eppley and Thomas (1969) obtained K_s values of 1.35 and 0.2 μ moles per liter of nitrate with *Asterionella japonica* and *Chaetoceros gracilis*, also in batch culture, at 26°C. K_s for nitrate uptake (which is not necessarily the same as K_s for growth on nitrate) was found by Eppley *et al.* (1969) to be positively correlated with cell size, with generation time, and with K_s for ammonium uptake. Oceanic species were found to have lower K_s values for uptake than inshore species, which usually grow in waters having higher nutrient concentrations.

The relationships of relative growth rate to nutrient concentration are, however, more complicated than the above simple hyperbolic expression of Monod suggests. The Michaelis-Menton kinetics which it represents may accurately describe uptake but relative growth rate is dependent more directly on intracellular concentration rather than the rate at which the nutrient enters the cell (see p. 41). The complications are particularly evident in the case of phosphorus (Fogg, 1973). This element appears to be taken up by algae in the form of orthophosphate only, but a variety of organic phosphates are utilized as a result of the activity of phosphatases which are produced at the cell surface, especially when inorganic phosphates are in short supply. Uptake of orthophosphate is an active process for which energy may be supplied by photosynthesis as well as by respiration and hence is usually stimulated by light. Given a supply of phosphate, algae are able to accumulate an excess which is stored within the cells in the form of polyphosphate (volutin) granules. The reserves resulting from this luxury consumption may then support growth in the absence of any further external supply. The freshwater diatom *Asterionella formosa* has been shown to be capable of accumulating sufficient phosphate reserves to provide for nearly seven doublings (Mackereth, 1953) and the marine diatom *Phaeodactylum tricornutum* may provide in a similar way for five successive doublings (Kuenzler and Ketchum, 1962). At certain stages in its division cycle *Chlorella fusca* excretes considerable amounts of organic phosphate (Soeder, 1970). The effects of these various complications are best studied by the technique of continuous culture and will be discussed further in the next chapter (p. 37).

Because phosphate and other nutrients may be accumulated in excess, the relative growth rate of an alga may not respond at once to a change in the external concentration of these nutrients. However, the response of exponentially growing *Asterionella formosa* to changes in factors such as light intensity and temperature is immediate (Lund, 1949). Thus, the view that cells acquire a certain growth potential which determines the rate of subsequent growth irrespective of changing environmental conditions, does not appear to be altogether correct.

It may also be noted that algae evidently possess considerable powers of adaptation to antimetabolites and antibiotics. By subculturing the blue-green alga *Anacystis nidulans* in gradually increasing concentrations of such substances as streptomycin, penicillin, chloramphenicol, sulfanilamide, and sodium selenate, Kumar (1964) produced strains resistant to their toxic effects. For example, in the course of 15 serial transfers he was able to produce a stable strain resistant to 50,000 times the maximum concentration of streptomycin tolerated by the original strain. Such adapted strains did not show greatly reduced relative growth rates in the presence of the antibiotic; k' for a strain resistant to 200 ppm of streptomycin and growing in medium containing that concentration was 0.347 as compared with 0.382 for the original strain in the absence of the antibiotic, both being measured at $33°$ C.

Sooner or later, exponential growth must cease in a culture of limited volume. The factors involved are various:

1. *Exhaustion of nutrients.* With cultures in the older types of media it is commonly the nitrate or iron supply which limits exponential growth. If this is so, then addition of further amounts of the limiting nutrient will prolong the exponential phase until some other factor becomes limiting. Unchelated ferric iron is precipitated as phosphate in alkaline medium; in this form it is largely unavailable to algae, and it is difficult to ensure that the supply is adequate. The introduction of chelating agents such as ethylenediamine tetraacetic acid (EDTA or versene) has enabled quantities of iron and other trace elements sufficient for prolonged exponential growth to be supplied in the medium without toxic effects.

Algae having a requirement for organic growth factors such as vitamin B_{12} may have the period of their exponential growth limited by the supply of these factors.

2. *Rate of supply of carbon dioxide or oxygen.* In stagnant cultures in a mineral medium the rate of diffusion of carbon dioxide into the culture from the air becomes limiting at a comparatively low population density. Improvement of the rate of aeration by shaking or stirring the culture or bubbling air through it will in this case prolong exponential growth. A supply of carbon dioxide enriched air may be necessary to maintain exponential growth in dense cultures. One or 5 per cent carbon dioxide is commonly supplied to algal cultures, but concentrations as high as this may have inhibitory effects, as, for example, with *Anabaena cylindrica* (Fogg and Than-Tun, 1960). Oxygen may similarly become limiting in heterotrophic cultures of algae, and here, again, aeration prolongs the exponential growth phase (Pearsall and Bengry, 1940).

3. *Alteration of pH of the medium as a result of preferential absorption of particular constituents of the medium.* It commonly happens that if nitrogen is supplied as an ammonium salt the preferential absorption of the ammonium ion causes the medium to become too acid to support growth. Absorption of nitrate ion results in an increase in *pH* but this is buffered by the medium's taking up more carbon dioxide so that it rarely affects growth to an appreciable extent. If carbon dioxide is limiting, the utilization of bicarbonate in photosynthesis may result in the *p*H of media rising as high as 11 or more, which may bring growth to an end. Utilization of organic acids without equivalent intake of cations may also result in the medium's becoming too alkaline for growth, as has been recorded for the colorless alga *Chilomonas* (Hutchens, 1948).

4. *Reduction of the light intensity by self-shading.* Light absorption by a *Chlorella* culture approximately follows Beer's law, the intensity of the penetrating light falling off exponentially as the path length through the algal suspension increases (Myers, 1953). This relationship is shown in Figure 11. It will be readily appreciated from the figure that, as a culture becomes dense, only the cells at the surface will receive a light intensity saturating for photosynthesis, the bulk of the culture being light-limited and, if the culture is very dense, in virtual darkness. In this situation growth is no longer determined by the size of

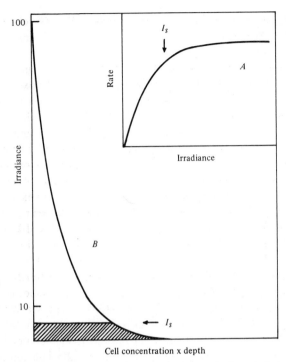

Figure 11. Light relations in a dense algal culture. A, the relation of rate of photosynthesis to irradiance, showing I_s, the estimated minimum irradiance for light saturation. B, decrease in irradiance in the culture as a function of cell concentration × depth. After J. Myers and J.-R. Graham, On the mass culture of algae. II. Yield as a function of cell concentration under continuous sunlight irradiance, *Pl. Physiol.* (1959) *34*:346, fig. 1.

the population but by the rate of light absorption. The growth curve therefore changes its character from exponential to linear, *i.e.*, growth becomes directly proportional to time, until some other limitation supervenes (Myers, 1953).

5. *Autoinhibition.* There is clear evidence that certain algae produce substances toxic to themselves in the course of their metabolism and that the accumulation of these may eventually bring exponential growth to a standstill. Such autoinhibition has been recorded for *Nostoc punctiforme* (Harder, 1917), a strain of *Chlorella vulgaris* (Pratt and Fong, 1940), and *Nitzschia palea* in impure culture (von Denffer,

1948; Jørgensen, 1956), but the substances concerned have not been fully characterized.

Since the changes which eventually bring exponential growth to a standstill begin as soon as a culture has been inoculated, it may occasion surprise that a protracted period of constant relative growth is ever possible. However, the changes in light intensity and in nutrient concentration brought about by growth are at first relatively small, and light and nutrients are normally supplied well above the limiting levels. From the nature of exponential growth it follows that the absolute amount of growth in any mean generation time is equal to the total of that in the period, however long, which has gone before. Hence, the reduction of the concentration of a nutrient from a level saturating for growth to zero is abrupt, and a gradual reduction in the relative rate of growth is consequently not manifest. Conversely, the production of an autoinhibitor is usually proportional to growth, so that its concentration remains at an innocuous level for a long period before suddenly mounting to the inhibitory level.

The duration of the period of declining relative growth depends on the nature of the limiting factor. Nutrient exhaustion or autoinhibition usually results in an abrupt transition from the exponential to the stationary phase, but, as already explained, if light is limiting, a prolonged phase of linear growth may intervene. Gaseous diffusion in stagnant culture is likewise dependent on the surface presented by the culture and, if limiting, produces a short phase of linear growth as found, for example, in chemotrophic cultures of *Chlorella* by Pearsall and Bengry (1940), before the decline into the stationary phase.

The final yield attained in the stationary phase depends on the nature of the limiting factor. If a nutrient is limiting, it is to be expected that the yield will be proportional to the amount supplied initially. This is the basis of the use of algae for the biological assay of substances such as vitamin B_{12} (Belser, 1963; Stein, 1973). Graded amounts of a standard solution of the vitamin are added to a medium which is known to support good growth of the test organism when supplemented with the vitamin. A parallel series of culture media with graded additions of the material to be assayed is also prepared. The cultures are inoculated with an organism having a specific requirement for the vitamin, *e.g.*, with *Euglena gracilis* or *Ochromonas malhamensis* when vitamin B_{12} is being determined, and population density is measured photometrically when the stationary phase is reached. From comparison of the response curves for the standard and the unknown, the concentration of the vitamin in the unknown may be estimated. Figure 12 shows a response curve for the marine flagellate *Monochrysis lutheri*. There is a linear relationship between final yield and vitamin B_{12} concentration up to 3 μg/liter (Droop, 1961*a*).

If autoinhibition has occurred, growth will finally cease when a particular concentration of cells is reached. In this case growth is resumed if the culture is simply diluted with distilled water or saline solution without further addition of nutrients (von Denffer, 1948).

By avoiding carbon, nitrogen, and iron deficiencies and ensuring maximum efficiency of light utilization by exposing the alga in layers of only 6 mm thickness, Myers *et al.* (1951) succeeded in obtaining

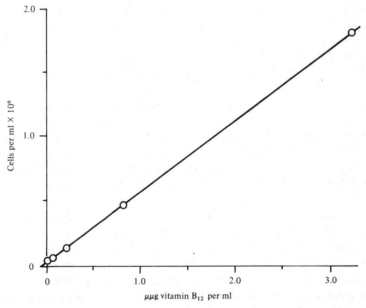

Figure 12. Relation of final yield of cells of *Monochrysis lutheri* to concentration of vitamin B_{12} in the medium. Cultures incubated at 15° C in a light intensity of 2,000 lux. From M. Droop, Vitamin B_{12} and marine ecology: the response of *Monochrysis lutheri, J. mar. biol. Ass. U.K.* (1961), *41*:73, fig. 3 (Cambridge University Press).

cultures of *Chlorella pyrenoidosa* with the extraordinarily high density of 54.9 gm dry weight of algal material per liter, or, in other terms, 25 per cent of the volume was occupied by algal cells.

Little can be said about the death phase. The time of its onset varies enormously according to species and the conditions of culture; often the stationary phase may be maintained for several weeks, but sometimes numbers may decline catastrophically immediately at the end of the exponential phase. This is so in cultures of *Ochromonas danica,* which normally show no stationary phase (Allen *et al.,* 1960). Lund (1959) has observed that illumination in the absence of sufficient amounts of certain nutrients results in breakdown of cellular organization and death in *Asterionella.* In bacterial cultures death normally follows an exponential course, the number of living cells, N_2, at time t_2, being related to the number N, at time t, by the expression

$$N_2 = N_1 e^{-k(t_2 - t_1)}.$$

Belcher and Miller (1960) found that the number of viable cells of *Monodus subterraneus* decreased in this way during incubation in the dark, and most simple algae probably show similar behavior.

A concluding comment in this chapter perhaps should be that there is no single criterion by which growth may be measured. The relative growth constant, k, the rate of growth in the linear phase, and the final yield are all useful measures but have different physiological and ecological significance and are affected differently by varying conditions. Miller and Fogg (1957) showed that the optimum concentrations and ratio of monovalent and divalent ions in the medium for exponential growth of *Monodus subterraneus* were different from those for final yield.

I I I

The Growth of Algae in Continuous and Synchronous Culture

For many experimental purposes batch cultures are decidedly unsatisfactory. Although conditions in cultures of this type may remain sufficiently the same to permit several days of constant relative growth, nevertheless, they are changing, and consequently growth is rarely "balanced," *i.e.,* attributes of the cells, such as mean size and composition, do not remain constant. Furthermore, any sample of the population will include cells in all stages of the division cycle, so that measurements made on it represent mean values, which give no clue as to the fluctuations that may be occurring during the growth and reproduction of the individual cells.

Continuous Culture

The method of continuous culture has been devised to overcome the first of these two difficulties by stabilizing the environmental conditions. Continuous culture consists essentially of holding a culture at some chosen point on its growth curve by the regulated addition of fresh medium. This can be done in a crude way by pouring off half of a culture after a period equal to the mean doubling time and making up the volume with fresh medium. A few workers, for example, Aach

(1952), have used this method, but obviously conditions in the culture may vary considerably in the time between the additions of medium, and unless the estimation of mean doubling time is accurate, it is difficult to hold the population at the desired level. However, Fay and Kulasooriya (1973) have described a simple and satisfactory continuous culture apparatus in which the addition of fresh medium is controlled manually.

Automatic continuous culture methods enable precise and continuous control of population density. Two different types of apparatus are available, the "turbidostat" and the "chemostat." The former, which was introduced by Myers and Clark (1944), has been most used in algal studies. In this apparatus, dilution is controlled by a photometric device to keep the population density, *i.e.*, the turbidity, of the culture constant and thus balance the rate of growth. In its original form the culture vessel consists of three concentric glass tubes, the outer pair providing a water jacket, the culture being held between the inner pair, with the space in the center being occupied by one of a pair of photocells (Fig. 13). Illumination is provided by fluorescent tubes arranged so that incident light is effectively independent of the volume of the culture. Filtered carbon dioxide enriched air bubbled through the culture serves to keep the cells in suspension. The second photocell faces the same light source as the other and is balanced against it by a neutral filter. Increased density of the culture throws the photocells out of balance and results in the opening of a solenoid valve to admit fresh medium. The culture can be maintained uncontaminated for two or three months, and portions are harvested manually at suitable intervals. Once a steady state is established, the conditions affecting growth remain constant: since the cell density remains constant, the average light intensity in the culture does so also; nutrient concentrations are maintained by the inflow of fresh medium, and products of metabolism accumulating in the medium are at the same time continually diluted.

Another apparatus employing the same principle has been described by Phillips and Myers (1954). This has a constant level overflow and has the advantages that it is easier to use and enables more precise measurement of relative growth rates. Maddux and Jones (1964) have devised a simpler apparatus with a long light path that permits continuous culture with as few as 28 cells of *Porphyridium cruentum* per mm^3 in suspension. Fuhs (1969) has also described a turbidostat unit.

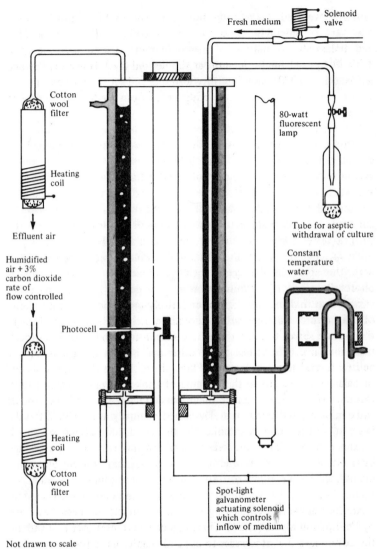

Fresh medium

Solenoid valve

Cotton wool filter

Heating coil

Effluent air

Humidified air + 3% carbon dioxide rate of flow controlled

Photocell

Heating coil

Cotton wool filter

80-watt fluorescent lamp

Tube for aseptic withdrawal of culture

Constant temperature water

Spot-light galvanometer actuating solenoid which controls inflow of medium

Not drawn to scale

Figure 13.   A diagrammatic cross section showing the general layout of a continuous culture apparatus operating on the turbidostat principle. Modified design after Myers and Clark, 1944.

In the turbidostat type of culture, growth is not limited by any nutrient but by factors intrinsic to the alga at the particular light intensity and temperature being used. The relative growth constant is given by an expression analogous to that which describes exponential growth in a culture of limited volume:

$$k' = \frac{\log_{10} V_t - \log_{10} V_o}{t},$$

where V_t and V_o are the total culture volumes at the end and beginning of the period of time, t. The cell populations produced have rather constant characteristics. For *Chlorella pyrenoidosa* grown at 25° C, samples taken at intervals over a period of three weeks had a mean value of k' of 0.564, with a maximum variation of 0.023, *i.e.*, about 4 per cent. Maximum apparent rate of photosynthesis was 0.687 mm³ of oxygen per minute per mm³ of cells, with a maximum variation of 0.05, *i.e.*, about 7 per cent (Myers and Clark, 1944). These variations are extremely small compared with those which occur during the course of growth in culture of limited volume.

Turbidostat culture has been used successfully for species of *Chlorella, Scenedesmus, Euglena, Anabaena,* and *Anacystis* (Myers, 1962a); *Monodus* (Miller and Fogg, 1957); *Porphyridium* (Jones in Oppenheimer, 1966); *Nitzschia* and *Tetraselmis* (Maddux and Jones, 1964); and *Cyclotella* and *Thalassiosira* (Fuhs, 1969). Organisms which stick to the glass walls of the vessel, *Platymonas subcordiformis* for example, cannot be grown by this method (Jones in Oppenheimer, 1966). Myers and his collaborators (Myers, 1946a, b; Phillips and Myers, 1954) applied the turbidostat technique in studies of the relation of photosynthetic and cellular characteristics of *Chlorella pyrenoidosa* to light intensity to which we have already referred (p. 25). Maddux and Jones (1964) found by means of turbidostat cultures that there is a pronounced interaction of nutrient concentration and light intensity in their effects on the relative growth rate of *Nitzschia closterium* and *Tetraselmis* sp. At low light intensities the concentrations of nutrients had little effect, but at medium light intensities, when the nitrate-nitrogen concentration was 140 mg per liter and the phosphorus concentration was 15.5 mg per liter, the maximum relative growth rate was about 4 times as great for *Tetraselmis* as it was when 0.125 and 0.012

mg per liter, respectively, of these nutrients were supplied. Further-more, growth took place at much higher light intensities when the nutrient concentrations were at the higher level than it did when they were at the lower level.

The other kind of continuous culture device, the "chemostat" (Monod, 1950; Novick and Szilard, 1950), has been used most exten-sively in the study of bacterial growth. Apparatus suitable for the growth of algae has been described by Droop (1966), Fuhs (1969), Caperon and Meyers (1972a, b), Fay and Kulasooriya (1973), and Stein (1973). It depends on the addition of fresh medium to the culture at a constant rate, the population density then adjusting itself to a maximum rate determined by the rate of supply of the limiting nutri-ent. Constant volume is maintained by an overflow device. The expres-sion given above, relating k' to culture volume, applies to chemostat as well as to turbidostat cultures. In this type of culture, a steady state is attained only if growth is limited by the rate of supply of a nutrient so that the relative growth rate must always be less than the maximum possible. A limiting situation, "wash-out," is reached when the rate of dilution of the culture by addition of fresh medium balances the increase of the population by growth. Fujimoto *et al.* (1956) deter-mined the relation of this limiting value of flow rate to light intensity for *Chlorella ellipsoidea.*

The chemostat is particularly suitable for the investigation of effects of nutrient concentration on growth. Droop (1968) used it in an important study of the kinetics of vitamin B_{12} uptake and effects on growth in *Monochrysis lutheri.* Uptake was followed using [57]Co-labeled vitamin B_{12} in exponentially growing cells in dilute culture. Ignoring a small amount of passive adsorption, uptake (u) was found to be related to concentration (c) according to the equation

$$u = \frac{u_\infty c}{c + K_c},$$

in which u_∞ is the maximum rate and K_c the half-saturation constant. This is a hyperbolic relationship of the same form as that for relative growth rate and nutrient concentration given on p. 29. Relative growth rate was related in a similar way to the concentration of vitamin within the cells (c_i)

$$k = \frac{k_\infty c_i}{c_i + K_s}$$

but has no direct relationship to the concentration in the medium. Droop's surmise that a similar situation might obtain with other nutrients has been borne out. Fuhs (1969) found the relative growth rates of two brackish water diatoms, *Cyclotella nana* and *Thalassiosira fluviatilis*, in chemostat culture to be related to the amount of bound phosphorus per cell rather than to the concentration in the medium. His data are in accordance with the view (p. 30) that restriction of the phosphorus supply affects mainly the storage of phosphorus in the cell and only severe limitation affects structural or functional components which determine the rate of growth. The same general picture holds for vitamin B_{12} with *Skeletonema costatum*, iron with *Monochrysis lutheri*, and nitrate with *Isochrysis galbana* (Droop, 1973). It thus appears that "luxury consumption" of nutrients is a general phenomenon and that consequently relative growth rates need bear no direct relation to contemporary concentrations of these nutrients in the medium. Where a relationship of the classical hyperbolic form has been found between relative growth rate and nutrient concentration, as in the studies of Soeder *et al.* (1971) on the effects of phosphate concentration on growth of *Nitzschia actinastroides*, the chemostat cultures used were allowed as long as 14 days to attain a steady state before readings were taken.

Droop (1973) pointed out that acceptance of Monod's equation for nutrient-limited growth implies acceptance of Liebig's law of the minimum in its unmodified form, that is, it is assumed that growth is limited only by the nutrient in shortest supply and that alterations in the concentrations of nutrients in more ample supply have little or no effects on the rate of growth. This is not in accord with observation, and Droop has proposed a polynomial extension of the Monod equation which has no discontinuity between the state of limitation and excess and allows for simultaneous limitation by more than one nutrient. An implication of this is that the yield of cell material per unit amount of nutrient supplied, such as may be determined by experiments of the sort represented in Figure 12, is not a constant but varies according to the levels of other factors.

Use of the chemostat has revealed other complexities in the relations of relative growth rate to substrate concentration. Droop (1968) found evidence of excretion by *Monochrysis lutheri* of a non-dialysable heat-labile inhibitor which combined with vitamin B_{12} , rendering it unavailable to the cells. This inhibitor was found to be produced by other species, including some not requiring vitamin B_{12} , and not to be species-specific in its effects. Soeder *et al.* (1971) found that in chemostat cultures of *Nitzschia actinastroides* there was a departure from the Monod model when concentrations of phosphate were very low, there actually being an increase, associated with the presence of dead cells, in phosphate concentration in the outflowing medium. The explanation seems to be that phosphate is excreted at certain stages in the life-cycle of the cell and this maintains or increases the concentration in the medium, enabling some cells to continue growth whilst others, which have excreted phosphate, pass below the critical level of cell phosphorus and die.

A type of culture intermediate between batch and continuous is that in which the population is grown in a vessel with a fine glass filter, or in dialysis tubing, on the other side of which the medium is constantly renewed (Jensen *et al.*, 1972). Concentrations of nutrients and extracellular products may be held constant by this means but the effective light intensity diminishes as the population increases.

Synchronous Culture

It is the aim of the method of synchronous culture to produce populations of cells uniform with respect to stage reached in the division cycle. The average properties of such populations are taken as representing the properties of individual cells as they grow in size and divide. This is a more satisfactory way to study the biochemical changes involved than making observations on individual cells. Techniques do exist for measuring the respiration rate, nucleic acid content, and various other attributes of individual cells, but these are generally difficult to apply and limited in scope.

Maaløe (1962), who has given a useful discussion of the rationale of synchronous culture with special reference to bacteria, has distinguished three types of technique for synchronizing cultures: (1) initial treatments, (2) repeated stimulus or entrainment methods, (3) mechan-

TABLE 4
Characteristics of "dark" and "light" cells of *Chlorella ellipsoidea* (from Tamiya *et al.*, 1953)

Cell type	Average cell diam. (μ)	Light-saturated photo-synthesis at 25° C	Light-limited photo-synthesis	Respiratory activity Q_{O_2} at 25° C	Chloro-phyll content (%)	Nitrogen content (%)
"Dark"	3.1–3.4	1.7 –1.9	0.36–054	4.6–6.1	2.4–5.2	7.0–9.5
"Light"	5.5–5.9	0.26–0.32	0.10–0.16	7.7–9.3	0.8–1.3	5.2–5.7

ical selection procedures. Tamiya, one of the pioneers in the field of synchronous cultures of microorganisms, at first used the third of these methods. By differential centrifugation he was able to separate two categories of cells primarily distinguished by size,* which he termed "dark" and "light" cells, from cultures of *Chlorella ellipsoidea* (Tamiya *et al.*, 1953). These cells had quite distinct characteristics, as shown in Table 4. In particular there was a contrast between the high photo-synthetic and low respiratory activity of the "dark" cells and the low photosynthetic and high respiratory activity of the "light" cells. By incubation of populations of these cells, the transformation of one type into the other could be followed. The principal features of the trans-formation of "light" into "dark" cells in the dark are shown in Figure 14. Since photosynthesis was impossible, there was no increase in cell material, and the total cell volume remained constant. The cells divided more or less together, cell numbers increasing fourfold, then remaining stationary, while there was a corresponding fall in mean cell volume to that characteristic of "dark" cells. These changes were dependent on the presence of oxygen. The growth of "dark" into "light" cells is light-dependent but does not require the presence of oxygen. The cycle is summarized in Figure 15.

Not only does the photosynthetic activity vary during the cell divi-sion cycle but there are evidently marked changes in the nature of the products. Nihei *et al.* (1954) produced *Chlorella* populations consisting

*It may be noted that an extremely simple sedimentation method for separating these two kinds of *Chlorella* cells for starting synchronous cultures has been described by Spektorov and Lin'kova (1962).

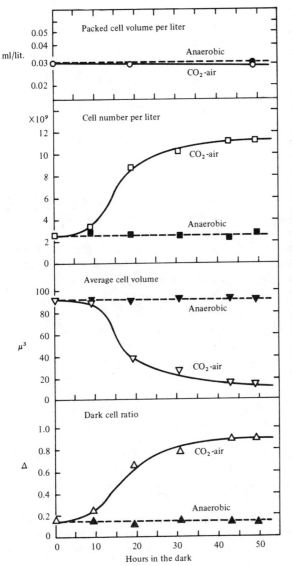

Figure 14. Change in packed cell volume, cell number, average cell volume, and dark cell ratio during the transformation of L cells of *Chlorella ellipsoidea* into D cells under aerobic and anaerobic conditions. Reproduced from H. Tamiya, T. Iwamura, K. Shibata, E. Hase, and T. Nihei, Correlation between photosynthesis and light-independent metabolism in the growth of *Chlorella*, *Biochim. biophys. Acta* (1953), 12:30, fig. 6 (Elsevier).

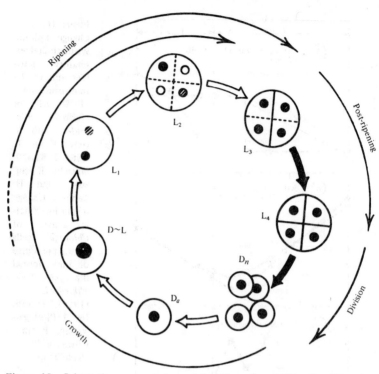

Figure 15. Schematic representation of the life cycle of *Chlorella ellipsoidea*. The white arrows denote light-dependent processes and the black arrows transformations occurring independently of light. The black dots denote Feulgen-staining nuclei. From H. Tamiya, Cell differentiation in *Chlorella, Symp. Soc. exp. Biol.* (1963), *17*:193, fig. 4 (Cambridge University Press).

of 90–95 per cent "dark" cells by incubating an actively growing culture in dim light for 7 days, that is, by method 1 above. Changes in activity occurring in such a population when it was transferred to bright light and became transformed to a population of "light" cells are shown in Figure 16. The photosynthetic quotient $(\Delta O_2/-\Delta CO_2)$ of "dark" cells was about unity, *i.e.*, that characteristic of carbohydrate synthesis, but rose to over 3, a value which suggests the production of highly reduced substances such as fats, when the transformation to "light" cells was complete. This supposition was borne out by elementary analyses of the cell material (Table 5).

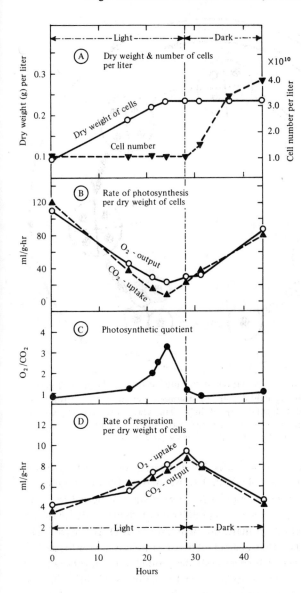

Figure 16. Change in photosynthetic and respiratory activities during the transformation of D cells of *Chlorella ellipsoidea* into L cells. From T. Nihei, T. Sasa, S. Miyachi, K. Suzuki, and H. Tamiya, Change of photosynthetic activity of *Chlorella* cells during the course of their normal life cycle, *Arch. Mikrobiol.* (1954), *21*:158, fig. 1 (Springer-Verlag, Berlin, Göttingen, Heidelberg).

TABLE 5

Elementary composition, in terms of percentage of dry material, of "dark" and "light" cells of *Chlorella ellipsoidea* (data of Nihei *et al.*, 1954)

Cell type	Carbon	Hydrogen	Oxygen	Nitrogen
"Dark"	39.3	6.7	37.5	8.1
"Light"	48.3	7.1	29.2	7.8

Using ^{14}C as a tracer, Kanazawa *et al.* (1970) estimated the rate of flow of carbon through various pools of metabolic intermediates at different stages in synchronized cultures of *C. pyrenoidosa*. The results, which showed a shift from amino acid and protein synthesis in the growth of "dark" cells to sucrose synthesis in predivision and divided cells (after darkness), are in general agreement with the above picture. These shifts were correlated with changes in activity of enzymes at several identifiable sites in photosynthetic carbon metabolism and related reactions.

It is now realized that the characteristics of "dark" and "light" cells, which Tamiya now prefers to call D and L cells, respectively, differ somewhat according to the previous treatment, and a number of subtypes has been recognized (Tamiya, 1964; see Fig. 15).

A serious uncertainty in studies with synchronous cultures is that it is difficult to decide whether observed changes are those normally accompanying the cell division cycle or whether they are induced by the treatment used to synchronize division (Pirson and Lorenzen, 1966), and there have been doubts as to whether the changes in rates of photosynthesis and respiration just described for the *Chlorella* cell cycle are typical. However, Sorokin (1964) separated cells of different age groups from a non-synchronized population of *Chlorella* by centrifugation and found that their photosynthetic characteristics corresponded with those of equivalent stages in cultures synchronized by light/dark cycles. Soeder and Ried (1967) studied respiration in synchronous cultures of *Chlorella* and concluded that the changes observed, which were similar to those described above, were correlated with the development of the cells and not directly dependent on the light/dark cycle to which the cultures were subjected. Senger (1970b), in studies with another green alga, *Scenedesmus obliquus,* found that both the quantum yield and photosynthetic capacity reached a maximum 8 hours after the beginning of the light period and then declined until the 16th hour, that is, until just before release of daughter cells. Thus, whereas Nihei *et al.* (1954) found a continuous decline in photosynthetic activity from stage D_n (see Fig. 15), Senger found an increase between stages approximating to D_n and D_a but there was otherwise agreement that D cells are more active than L cells. From a review of published results, Senger (1970*a*) concluded that this change in photosynthetic capacity is inherent in the normal life cycle of simple algae.

Other work with synchronous cultures has been concerned with the biochemistry of cell division. This has been reviewed by Tamiya (1964, 1966) and need not be discussed at length here, but the effects of deficiency of certain nutrients and inhibitors on division are of some interest. In media lacking nitrate, phosphate, potassium, or magnesium, D_a cells can perform one cycle of division, giving rise to different numbers of daughter cells, which are then unable to develop normally. In a sulfur-deficient medium, however, cells grow only to the stage of early ripening (Fig. 15), then cease growing and do not divide. A similar effect is produced by antagonists of sulfur-containing amino acids, such as ethionine and allylglycine. A certain competition between growth and cell division has been observed, for L_2 cells (Fig. 15) continue to grow if kept in the light, although they are capable of limited division if transferred to the dark. It seems that protein synthesis and synthesis of specific sulfur-containing peptide-nucleotide substances necessary for cell division compete for supplies of nitrogen and sulfur. Under photosynthetic conditions protein synthesis predominates and thus cell division is prevented if these supplies are limited (Hase, 1962; Tamiya, 1963).

Continuous cultures may be synchronized by imposition of appropriate light/dark cycles. Researchers involving this have been described by Paasche (1967, 1968), Gimmler *et al.* (1969), and Eppley *et al.* (1971). These and other studies on the effects of photoperiod have revealed a variety of response amongst phytoplankton species. Whereas *Coccolithus huxleyi* in synchronized cultures divided in the dark (Paasche, 1967), cell division in *Ditylum brightwellii* and *Nitzschia turgidula* took place in the light (Paasche, 1968). Quraishi and Spencer (1971), who used an apparatus giving continuously varying light intensity similar to that found under natural conditions rather than an abrupt transition

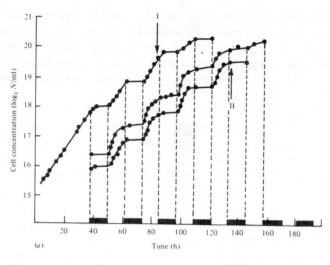

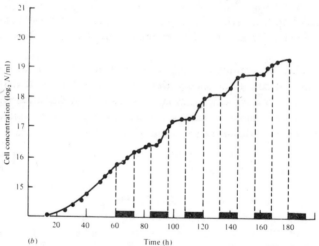

Figure 17. Induction of rhythmic growth in phytoplankton species on transfer from continuous illumination to a regime of 12 h dark and 12 h light of intensity 6000 lux. The bars enclosed by the broken lines indicate the dark periods. a, *Phaeodactylum tricornutum;* b, *Brachiomonas submarina.* From F. O. Quraishi and C. P. Spencer, Studies on the responses of marine phytoplankton to light fields of varying intensity, in *Fourth European Marine Biology Symposium,* ed. D. J. Crisp (1971), pp. 393–408, fig. 2 (Cambridge University Press). The arrows I and II indicate times of subculturing.

between light and dark, made the observation that among the five species which they studied, *Phaeodactylum tricornutum* and *Chlorella ovalis* divided during light periods whereas *Brachiomonas submarina*, *Dunaliella primolecta*, and *Monchrysis lutheri* preferentially divide in the dark (Fig. 17). *Chlamydomonas reinhardi* also preferentially divides in the dark (Mihara and Hase, 1971).

Although the growth rate of *Chlorella ellipsoidea* is almost proportional to length of photoperiod all the way up to continuous light (Tamiya *et al.*, 1955) in *Coccolithus huxleyi*, *Ditylum brightwellii*, and *Nitzschia turgidula*, growth in continuous light was never significantly greater than it was with 16 hours light and 8 hours dark (Paasche, 1967, 1968). The colonial green alga *Hydrodictyon* seems, however, to be one of the few algae which are unable to grow unless illumination is interrupted by dark periods (Pirson, 1957). Eppley *et al.* (1971) found that *Skeletonema costatum* assimilated nitrate and ammonium nitrogen mainly during the light and at a lesser rate during the dark. *Coccolithus huxlyei* assimilated both nitrogen sources at a high rate in both light and dark but the activity of nitrogen-assimilating enzymes was higher in cell-free extracts made from illuminated cultures than from those in the dark.

In studies on continuous cultures of *Skeletonema costatum* under conditions of silicate deficiency Davis *et al.* (1973) found that sexual reproduction became synchronized. This enabled a detailed description of the life cycle to be obtained, and it was pointed out that continuous culture offers a valuable means of studying the life cycles of planktonic algae.

Metabolic Patterns and Growth

From what has already been said, it will have been evident that considerable variability in metabolic activity and in the products of metabolism is encountered when algae are grown in laboratory culture. Since similar variability presumably occurs in natural environments, it is important to take this into account in interpreting the behavior of phytoplankton, and it is worth considering in more detail the relation of metabolic pattern to growth.

An indication of the extent of variation possible in metabolism is given in striking visual form if young and old cultures of *Botryococcus braunii* are compared. The young colonies show the characteristic green pigmentation of their group, the Chlorophyceae, and sediment to the bottom of the flask. By contrast, the colonies in old cultures, in which nitrate is exhausted, accumulate carotenoid pigments, which completely mask the chlorophyll, and lipoids, which lower the specific gravity of the colonies so that they float. Equally striking quantitative differences are shown by analyses of the cell material of *Chlorella* and other algae (for references see Fogg, 1959). Those summarized in Figure 18 are for *Monodus subterraneus*. When growing exponentially, this alga has a protein content of almost 70 per cent of the dry weight, high contents of chlorophyll and nucleic acids, low contents of carbohydrate and fat, and high photosynthetic and respiratory activity. The larger cells from an old, nitrogen-deficient culture, on the other hand,

52

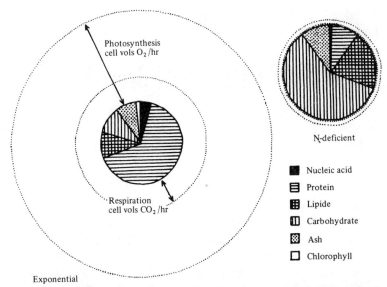

Figure 18. Diagram showing the differences in rates of metabolism and cell composition between exponentially growing and nitrogen-deficient cells of *Monodus subterraneus*. The central areas show the relative cell diameters and compositions on a dry weight basis of the cell material. The outer annular areas are proportional to the rates of respiration and photosynthesis respectively. Data of Fogg, 1959.

have a protein content of less than 10 per cent, low contents of chlorophyll and nucleic acids, high contents of reserve polysaccharide and fat, and extremely low photosynthetic and respiratory activity.

These, of course, are mean results for large unsynchronized populations of cells, and to some extent the differences must be due to the different proportions of cell types present at different stages in the growth of the cultures. Thus, there are certain resemblances between the characteristics of "average" cells from exponentially growing cultures and those of D cells as described by Tamiya; both are small, active in photosynthesis, and with high cell nitrogen content. There is also a resemblance between "average" cells from old cultures and Tamiya's L cells, both being large, inactive in photosynthesis, and low in nitrogen. However, it does not seem that the variation in average cell properties observed during the course of growth of cultures of limited volume can be entirely accounted for in terms of variation in proportions of cells in different stages of the "normal" cell cycle. There is, for example, the important difference that L cells have a high rate of respiration, whereas cells from an old culture have low respiratory activity. Furthermore, the variations during the course of growth in culture are more extreme than those which have been recorded for synchronous cultures.

There have been few studies directly concerned with the metabolism of algal cells in exponential growth, for the simple reason that the biochemist avoids the complication of increase in cell material in his experiments unless he is actually studying growth processes. It is clear, however, that exponentially growing cells of algae have high photosynthetic capacity and that the main product of their photosynthesis is protein. Efficiencies of up to 25 per cent in conversion of radiant energy into potential chemical energy have been observed in actively growing cultures of *Chlorella* (Wassink, 1954). Corresponding with the high proportion of protein found by direct analysis in such actively growing cells, the photosynthetic quotient under these conditions is greater than unity (Myers, 1962*b*). Carbon fixed in photosynthesis is known to be incorporated rapidly into amino acids (Bassham and Calvin, 1957), and it has been found that in actively growing cells of *Navicula pelliculosa* a high proportion of carbon fixed photosynthetically appears within 30 sec in a cell fraction containing the protein (Fogg, 1956). Protein may thus be regarded as the main and direct product of photosynthesis in this phase of growth. It may be noted that the relative rate of growth bears no direct relation to the total nitrogen content of the cells (Fogg, 1959). As with bacteria (Maaløe, 1962), it is likely that the controlling factor here is a ribonucleic acid fraction.

The photosynthetic activity of cells changes appreciably in the course of the exponential phase in batch culture. With the marine diatoms *Phaeodactylum tricornutum, Nitzschia closterium,* and *Chaetoceros* sp., Ebata and Fujita (1971) observed rates of light-saturated photosynthesis per cell to increase to a maximum at the early or middle exponential phase. A rapid decrease in activity occurred during the late exponential phase and a minimum was reached in the stationary phase. Morris and Glover (1974), who found similar changes with other marine phytoplankton, observed that the time of the maximum in photosynthesis depends on the temperature at which the cultures are grown. They point out that this invalidates earlier work on temperature adaptation (Steemann Nielsen and Jørgensen, 1968) in which comparisons were made between samples taken after the same period of growth in batch cultures at different temperatures. Clearly, in making such comparisons, it is necessary to use material in similar physiological states and continuous culture offers the best means of ensuring this.

The effects of transferring exponentially growing cells to a medium lacking a source of assimilable nitrogen are of interest from both the physiological and the ecological points of view. Cells treated in this way retain their high capacity for photosynthesis for several hours, but in the absence of a nitrogen source the product must necessarily be different. The flow of carbon fixed in photosynthesis is switched from the path of protein synthesis to that leading to carbohydrate, as shown by the evidence of the photosynthetic quotient, direct analysis (for references see Fogg, 1959), and tracer experiments (Holm-Hansen *et al.*, 1959). As a result of this continued photosynthesis the dry weight of cell material in the nitrogen-starved suspension increases. Cell division takes place at first but does not continue after the cell nitrogen has fallen below a limiting value, which is generally of the order of 3 per cent of the dry weight. Thereafter continued photosynthesis results in cell expansion and increase in mean dry weight per cell.

On prolonged nitrogen starvation (Fig. 19) the maximum rate of photosynthesis of which the cells are capable diminishes, eventually becoming only 5 per cent or so of the initial rate. This is accompanied by a decrease in the chlorophyll content of the cells, but the decrease is evidently not the only cause of the impairment of photosynthetic capacity (Ebata and Fujita, 1971). Respiration likewise falls off but after reaching a minimum shows a rise, which may be surmised to indicate the beginning of breakdown of cell organization. From this point onwards fat begins to accumulate in the cells. Fats do not require nitrogen for their synthesis but this may also be the result of the fat-synthesizing enzyme system's being less susceptible to disorganization than is the system responsible for carbohydrate synthesis, so that the fat-synthesizing enzyme system now gets the major share of the carbon fixed in photosynthesis (Fogg, 1959).

It cannot be assumed that the changes occurring in cultures which exhaust their nitrogen in the course of growth are necessarily similar to those just described for exponentially growing cells transferred to a medium lacking a nitrogen source. In the former case the cells are

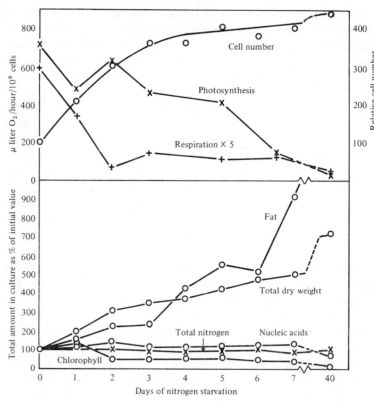

Figure 19. The effects of transferring exponentially growing *Monodus subter-raneus* to a medium containing no source of combined nitrogen. Data of Fogg, 1959.

subjected to falling concentrations of other nutrients and rising con-centrations of by-products of metabolism as well as to nitrogen defi-ciency, and these may well exert their effects on the pattern of metabolism.

In ageing cultures of *Chlorella* there is a marked decrease in photo-synthetic capacity. Van Oorschot (1955) found that the quantum efficiency remained constant at about 0.07 mol O_2 quantum so long as nitrate was present, but that efficiency fell, when nitrogen was ex-hausted, to about 0.02 mol O_2 quantum within 24 hours. Provided that the culture was not allowed to continue for more than a few days

without a supply of nitrate, the original quantum efficiency was re-
stored within 24 hours after addition of further nitrate. In this instance
the decline in photosynthetic capacity seems directly related to exhaustion of the nitrogen supply, although Pratt (1943) found that metabolic
products inhibiting photosynthesis accumulated in the medium in
which another strain of *Chlorella* was grown.

Many sets of analyses (for references see Fogg, 1959) show that fat
accumulates in large quantities in nitrogen-deficient cultures of Chlorophyceae, Xanthophyceae, and diatoms. Corresponding with this, the
distribution of radioactivity among various cell fractions of *Navicula
pelliculosa* after photosynthesis for 2 minutes in the presence of ^{14}C-
labeled bicarbonate has been found to be markedly different in material
from old, nitrogen-deficient cultures from that in material in actively
growing cultures, as much as 70 per cent entering the fat fraction in the
old cultures, as contrasted with less than 20 per cent in the latter case
(Fogg, 1956). There is no evidence of a preliminary phase of carbohydrate accumulation such as is found in actively growing material
abruptly transferred to nitrogen-deficient conditions. This seems ex-
plicable in terms of the different cell types which would be expected to
predominat~   ~ditions. An actively growing population
consi·   ~n deficiency does not markedly affect
t'   ~ation o.   ) L cells, and the L cells produced can
    division. The second generation of D cells, however, is unable
᠊evelop normally, and it seems that the enzymic composition of
᠊nese cells may then be fixed in the state characteristic of this phase,
namely, one which favors the synthesis of carbohydrates (see p. 46). On
the other hand, cells in a culture in which the nitrogen supply becomes
exhausted as a result of growth become arrested at the L stage,
presumably as a result of the greater sensitivity of the L–D transformation to inhibitory products of metabolism under conditions of nitrogen
deficiency. These cells characteristically form highly reduced products
of photosynthesis (see p. 46), and fat accumulation is evidently a result
of this tendency. It may be noted that actively growing cells transferred
to conditions of sulfur deficiency are halted in the L stage, not in the D
·tage as with nitrogen-deprived cells, and these accumulate fat (Otsuka,
᠊1).

        ᠊eems to be no particular effect of nitrogen deficiency on the
        ᠊f the miscellaneous fat-solvent soluble substances classed

as lipoids. Collyer and Fogg (1955) found that the proportion of this fraction in a number of species showed no correlation with fatty acid content or inverse correlation with cell nitrogen content.

Accumulation of carotenoid pigments, a characteristic of many algae when growing under natural conditions, often occurs, however, in ageing cultures in which nitrate has been exhausted. Even if sufficient carotenoid to make the cells bright orange or red is not produced, an increase in the carotenoid/chlorophyll ratio is usually detectable in such cultures (Fogg, 1959). A study by Droop (1954) with *Haematococcus pluvialis* showed that the accumulation of carotenoid is favored by circumstances, including phosphate deficiency as well as nitrogen deficiency, which prevent cell division without impairing the alga's ability to assimilate carbon.

Few studies have been made of the effects on patterns of metabolism of deficiencies of nutrients other than nitrogen. The rapid decrease in photosynthetic activity, referred to above, in cultures of three species of marine diatoms growing in an artificial sea-water medium, has been attributed to phosphate deficiency as the main cause (Ebata and Fujita, 1971). Ketchum *et al.* (1958) observed that in *Dunaliella euchlora* the ratio of net to gross photosynthesis fell sharply in phosphorus-deficient cells. This implies a similar climacteric rise in respiration as observed with nitrogen deficiency.

The effects of nutrient deficiency on the relationship between rate of photosynthesis at light saturation and chlorophyll content deserves some comment since they are important for the interpretation of the behavior of natural populations. The relationship is expressed in terms of the assimilation number (= mg C fixed per hr per mg chlorophyll *a* at light saturation). Spoehr and Milner (1949) found that assimilation number rose in nitrogen-deficient *Chlorella*. Likewise, Fogg (1959) observed a value of 6.1 in nitrogen-deficient *Monodus* as compared with 1.6 in actively growing cells. However, falls in assimilation number in nitrogen-deficient cultures have been recorded by Bongers (1956) for *Scenedesmus* and by McAllister *et al.* (1964) for *Dunaliella* and *Skeletonema*. The tropical marine diatom *Chaetoceros gracilis* grown in nitrogen-limited chemostat culture was found to show an increase in assimilation number with increasing cell nitrogen content and increasing growth rate (Thomas and Dodson, 1972). The causes of these contradictions, which are also evident in observations on natural populations, are probably numerous. The light-saturated rate of photosynthesis depends on the activities of enzymes which bear no simple relationship to the amount of chlorophyll in a cell and which are affected differently from chlorophyll by nutrient deficiency. The cells studied by Spoehr and Milner (1949) and Fogg (1959) were obtained by nitrogen starvation for long periods and were certainly in a different physiological state from those used by Bongers (1956) and McAllister et al. (1964), which were subjected to much shorter periods of deficiency, and from those in Thomas and Dodson's (1972) continuous cultures, which were actively growing. There is no reason to suppose that cells of one species, let alone of different species, in these diverse physiological states should show similar alterations in photosynthetic enzymes in relation to chlorophyll when subjected to nutrient deficiency. The conditions under which assimilation number is measured may also affect its value; sometimes excessively high light intensities are used, which may cause reduced photosynthetic yield by increasing photorespiration or may cause photo-oxidation of chlorophyll. The only conclusion, pending a more detailed knowledge of algal metabolism than we have at present, must be that assimilation numbers should be interpreted with caution.

Different species of algae have a tendency to resemble each other in the relative amounts of crude protein, fats, and hydrolysable polysaccharide which they contain when grown under approximately similar conditions (Collyer and Fogg, 1955). There may be some differences between algal classes in respect of general composition, but, with certain exceptions to be mentioned below, these are small compared with the differences in cell composition which a single species may show in the course of growth in culture. Parsons et al. (1961) analyzed eleven different species of marine plankton representing 6 algal groups, all grown under similar physical and chemical conditions and harvested in the exponential phase. The ash content of the different species varied greatly, being especially high in the diatoms, but if allowance is made for this by expressing the amounts of major fractions in terms of the total organic carbon in the cells, then the composition of the cells is generally similar (Table 6). It will be seen that the maximum variation in protein content here is only a little more than twofold, whereas in *Chlorella pyrenoidosa* the variation may be more than tenfold, according to the conditions under which it is grown (Spoehr and Milner,

TABLE 6

Comparison of the cell composition of species of marine phytoplankton, grown under similar chemical and physical conditions, in terms of the ratio of components to (oxidizable) carbon (Parsons *et al.*, 1961)

	Protein/C	Carbohydrate/C	Fat/C
Chlorophyceae			
Tetraselmis maculata	1.42	0.41	0.07
Dunaliella salina	1.43	0.80	0.15
Chrysophyceae			
Monochrysis lutheri	0.94	0.59	0.22
Haptophyceae			
Cricosphaera (Syracosphaera) carterae	1.41	0.45	0.12
Bacillariophyceae			
Chaetoceros sp.	1.12	0.22	0.21
Skeletonema costatum	1.38	0.79	0.17
Coscinodiscus sp.	1.08	0.27	0.11
Phaeodactylum tricornutum	0.88	0.64	0.17
Dinophyceae			
Amphidinium carteri	0.69	0.75	0.44
Exuviella sp.	0.70	0.84	0.34
Cyanophyeae			
Agmenellum quadruplicatum	0.86	0.75	0.31

1949). The low values for the protein/C ratio in the representatives of the Dinophyceae are probably attributable to the thick cellulose walls which they possess. If this were taken into account, their cell composition would conform more closely to that of the others.

An exception to the generalization that the gross composition of the cell material of an alga is dependent more on environmental circumstances than on class or species seems to be provided by members of the Rhodophyceae and Cyanophyceae. These, unlike the algae discussed above, do not appear to accumulate fat under conditions of nitrogen deficiency (Collyer and Fogg, 1955). This may be related to the finding of Erwin and Bloch (1964) that the mechanism for biosynthesis of unsaturated fatty acids in the Rhodophyceae and Cyanophyceae is different from that in the Chlorophyceae and higher plants.

Of course, algae may differ considerably in details of composition even when differences due to environmental conditions are eliminated, and it is firmly established that the various algal groups are characterized by particular pigments, carbohydrates, sterols, and other compounds.

V

The General Features of Phytoplankton Growth in Lakes and the Sea

It is quite beyond the scope of this book to review the mass of data relating to phytoplankton which limnologists and oceanographers have been accumulating since quantitative observations were first begun by Hensen in 1887. In fact, much of this information is of little value for the present purpose of describing the general patterns of phytoplankton growth and relating the salient features to the findings of laboratory studies. Techniques for quantitative estimation of phytoplankton, a necessary prerequisite, of course, for the accurate following of its population dynamics, are far from satisfactory. Details of methods are to be found in reviews such as those of Lund and Talling (1957), Strickland and Parsons (1968), Strickland (1972), and Vollenweider (1974), but some discussion of the chief difficulties should be given here.

Both horizontally and vertically the distribution of phytoplankton is usually patchy. The taking of samples at successive depths is normal limnological and oceanographic practice, but even samples taken at such relatively close intervals as one meter may be inadequate to give an accurate estimate of abundance; divers occasionally report seeing thin horizontal laminae of organisms, only a centimeter or so in thickness. Problems of microstratification have been discussed by Cassie (1963), and a sampler for thin horizontal layers has been described by Parker *et al.* (1968). A device, such as a pipe (Lund and Talling, 1957), which samples the whole of a water column, is more satisfactory for obtaining an integrated sample than is the pooling of a number of samples taken with the usual type of water-sampling bottle. Heterogeneous horizontal distribution is more difficult to take into account. That blooms of phytoplankton may be extremely patchy is sometimes very obvious as one flies over coastal waters or a lake. Patches of marine plankton, which are usually elliptical in shape, vary in a continuous series from a few feet across to as much as 30 or 40 miles by 120 or 180 miles, the mean being about 10 by 40 miles (Bainbridge, 1957). For the larger and more robust species the continuous plankton recorder of Hardy (1956) may provide a record of the major features of distribution in the sea but not of small-scale variations. Long narrow bands or streaks, a few centimeters or a meter or so in width, commonly forming a pattern superimposed on that of the patches, are produced by Langmuir circulation (see p. 72). Rodhe (1958) showed how samples taken at a single station on a lake vary from day to day, with wide fluctuations related to wind-induced water movements. Highest population densities were correlated with winds blowing from the direction of the shallow, most fertile part. Species which swim or float to the surface may be concentrated against a lee shore or in regions of downwelling, giving rise to the dense accumulations in the sea which go by the name of "red tides." Information about exceptionally dense populations in the sea has been summarized by Hart (1966). Clearly, many sampling stations distributed over a wide area, or some means of following and sampling a specific water mass, are needed if phytoplankton growth is to be followed under these circumstances. Verduin (1951) compared phytoplankton data obtained by multiple mobile sampling with those from a single station, in western Lake Erie. However, the pattern of distribution may be so complex as to defeat even the most intensive attempts to obtain a picture by means of samples taken at discrete stations or along transects. In a situation such as that illustrated in Plate 1 a remote sensing technique seems to be the only possibility (Horne and Wrigley, 1974).

Much information relates to samples obtained by means of nets. This method has the obvious advantages of increasing the volume of water sampled and giving an integrated sample representing a large area. It is,

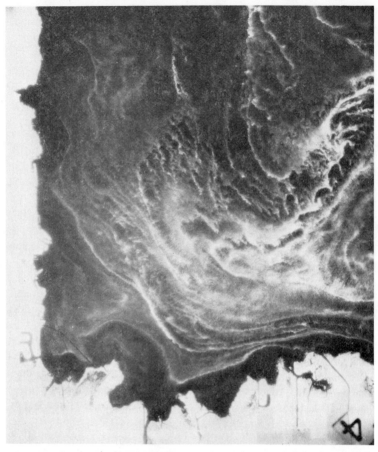

Plate 1. A bloom of blue-green algae in Clear Lake, California, photographed from the air. This infra-red (730-950 nm) image of the southwest corner of the lake was taken on May 11, 1973, from an altitude of 17,000 feet above mean sea level. The area represented is about 4.5 km across. The nearly straight lines traversing the bloom, *e.g.*, in the bottom left corner, are the tracks of boats. By courtesy of R. C. Wrigley (NASA, Ames Research Center) and A. J. Horne (University of California, Berkeley).

however, clear that even the finest net misses an important part of the phytoplankton. Verduin (1956) measured photosynthesis in lake water before and after filtration through the finest bolting silk (64 μm apertures) and found, on the average, 65 per cent of the activity in the filtrate. Rodhe (1958) likewise found that the rate of photosynthesis in lakewater samples was more closely correlated with the numbers of nannoplankton than with the numbers of plankton retained by a plankton net. On the basis of direct counts, Lund (1961) estimated that "μ-algae" generally account for less than 10 per cent of the total algal mass in the English lakes. Evidently, as one would expect from their high surface/volume ratio, these small forms are more active per unit mass of cell material than the algae retained by a net. With seawater samples Holmes and Anderson (1963) found that more than half the photosynthesis was carried out by algae which passed through a net with 35 μm apertures. Platt and Subba Rao (1973) also emphasized the importance of the contribution of the nannoplankton (which they define as being between 5 and 65 μm in diameter) to primary production in the sea. Reynolds (1973), who used a fluorimetric method for the determination of ultraplankton (defined as being less than 15 μm in diameter), concluded that in the Barents Sea and some other northern waters it may sometimes contribute over 90 per cent of the total chlorophyll a in the water. Mommaerts (1973) investigated primary productivity in the South Bight of the North Sea and, in spite of the water column being well mixed at the time, found that the net−/ nanno- plankton ratio varied markedly in the horizontal direction, with net-plankton photosynthesis being dominant near the coast and nannoplankton photosynthesis greater further out to sea.

It is thus clear that net sampling is inadequate for quantitative work. Collection of large volumes of water and separation of the algae by sedimentation, centrifugation, or filtration through a fine filter seem to be the best methods. Even these are not entirely satisfactory, since certain species are extremely fragile and break up when filtered or treated with chemical preservatives (Reynolds, 1973). For such species, counts must be made on fresh samples within a few minutes of collection, photomicrographs being taken for later identification of unfamiliar forms. Preservation of some of these delicate organisms is possible with the newer fixatives developed for the purposes of electron microscopy. Quantitative culture methods may be employed for determining occurrence and abundance but many nannoplankton species are difficult to grow in culture.

The total biomass of phytoplankton may be determined in terms of volume, carbon content, chlorophyll content, or various other measures. Chlorophyll is the most commonly used measure. It may be determined down to 1 mg per m^3 by absorption spectrophotometry (Strickland and Parsons, 1968; Stein, 1973) or to a fiftieth of this amount by fluorimetry (Holm-Hansen et al., 1965; Reynolds, 1973). One of the best chemical indicators of biomass is perhaps adenosine triphosphate since this is only found in nature in living cells and can be estimated by the extremely sensitive luciferin-luciferase "firefly" reaction and a photomultiplier (Holm-Hansen, 1971; Strickland, 1972). From what was said in the previous chapter it will be realized that any of these measures is likely to vary widely in relation to the others and expressions, as for example those by which cell carbon is estimated from volume (Reid et al., 1970; Strickland, 1972), can only be regarded as approximate. Chlorophyll fluorescence in natural phytoplankton populations, besides undergoing a photoinhibition near the water surface, shows a diurnal variation with a maximum value up to four times the minimum, as well as other, less predictable, variations (Kiefer, 1973).

Recognition and counting of numbers of individual species is laborious but undoubtedly provides the most useful kind of information. Electronic dimensional particle counters (Coulter counters), which have been used successfully with pure cultures, must be used with great caution on natural populations, which may contain as much as 80 per cent of particulate matter in the form of detritus as well as organisms of different shapes and sizes (Stein, 1973). Given the mean dimensions of the cells, counts may be converted to volumes or to cell surface areas. Paasche (1960a) studied the relation of rates of photosynthesis, as determined by the ^{14}C method, to standing crop, expressed in terms of numbers, volumes, and cell surface area, in samples from the Norwegian Sea. As Figure 20 shows, these three measures of standing crop give strikingly different impressions of the importance of the various species. Total cell surface area was most highly correlated with rate of photosynthesis, the correlation coefficient between the two being 0.74, as compared with 0.45 and 0.62, for cell number and cell volume, respectively. Possibly, then, cell surface area is the most adequate

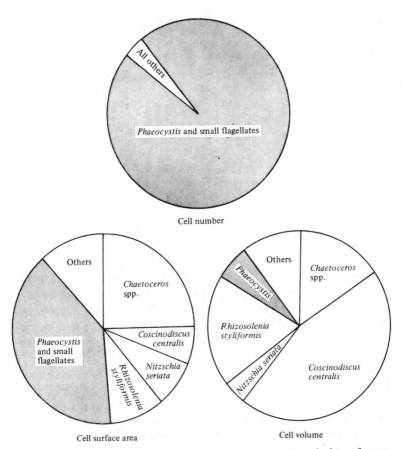

Figure 20. Relative proportions of various species of phytoplankton from a station in the Norwegian Sea expressed in terms of cell numbers, total cell surface area, and total cell volume. From E. Paasche, On the relationship between primary production and standing stock of phytoplankton, J. *Cons. perm. int. Explor. Mer* (1960), *26(1)*:42, fig. 6.

measure of the standing stock of phytoplankton. When individual species, rather than total phytoplankton, are being studied, the various possible measures may reasonably be assumed to be sufficiently well correlated as to give equally useful pictures of abundance.

While these various sources of uncertainty must be borne in mind in considering details, the general features of phytoplankton growth are clear enough. There is first of all a fairly well-defined seasonal periodicity in total biomass. In temperate and polar waters, both fresh and salt, there is no appreciable growth during the winter. Phytoplankton numbers increase early in the spring, generally reaching a maximum toward the end of April. So abrupt and rapid is this increase that British biologists usually refer to it as the spring outburst. The peak is followed by an almost equally steep decline, and during the summer numbers remain at a relatively low level. A second maximum, usually not so great as the one in the spring, may occur in the autumn, after which the numbers decrease to the low winter level. This pattern is shown both by the phytoplankton as a whole (see Fig. 21) and, sometimes, by individual species such as *Asterionella* (Fig. 22).

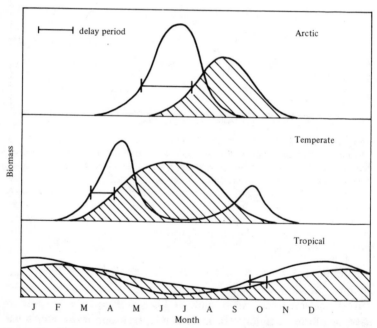

Figure 21. The seasonal variations in phytoplankton and herbivorous zooplankton (hatched area) in different latitudes. The horizontal bar indicates the delay period between the increases of phytoplankton and zooplankton. From D. H. Cushing, The seasonal variation in oceanic production as a problem in population dynamics, *J. Cons. perm. int. Explor. Mer* (1959), 24:455–464, fig. 3 (Høst and Son, Copenhagen).

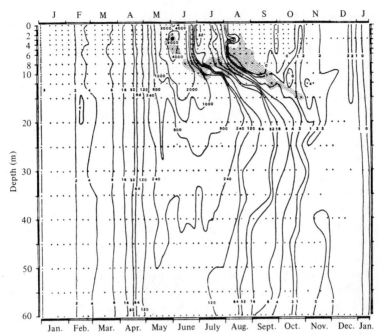

Figure 22. The distribution in depth and time of numbers of live cells of *Asterionella formosa* per ml in Windermere, North Basin, during 1947. The dots show the depths at which samples were taken for the estimations. Isopleths (lines of equal numbers) have been drawn freehand. The stippled area shows the depth-time distribution of the greatest vertical density gradients (the pycnocline), associated with the thermocline. From J. W. G. Lund, F. J. H. Mackereth, and C. H. Mortimer, Changes in depth and time of certain chemical and physical conditions and of the standing crop of *Asterionella formosa* Hass. in the north basin of Windermere in 1947, *Phil. Trans. R. Soc., B* (1963), *246*:265, fig. 3.

In Arctic and Antarctic waters there is a single peak at about mid-summer, generally similar to the spring maximum in temperate waters in its principal features. An example of this type of periodicity has been given by Horne *et al.* (1969). In both temperate and polar waters the seasonal amplitude in phytoplankton numbers is great, of the order of a thousandfold. By contrast, in tropical waters the seasonal variation may be as little as fivefold (Fig. 21), as, for example, Qasim *et al.* (1969) found in an estuary on the west coast of India.

These seasonal pictures, which, of course, grade into each other, apply to most lakes and seas but may sometimes be modified by local conditions. For example, where algal production is dependent on the upwelling of cold nutrient-rich water, as in the Benguela and Peru currents, the peak in algal growth is determined by the season of upwelling. Production in tropical lakes may reach a maximum following an influx of nutrient-rich water in the rainy season (Holsinger, 1955). Algal increase in Indian coastal waters is largely determined by mixing in of nutrient-rich water from below during the monsoons (Shah, 1973).

Phytoplankton shows qualitative changes with season as well as variation in total biomass, there being a seasonal succession of species occurring fairly consistently from year to year in a given lake or area of sea. In addition, there are differences in spatial distribution, although in comparison with flowering plants phytoplankton species are characteristically cosmopolitan in distribution. Some tendencies towards different geographical distributions are shown by the phytoplankton groups. Thus although diatoms are abundant in all oceans they are usually dominant in Antarctic waters, whereas dinoflagellates are more prominent in tropical waters and the important genus *Ceratium* seems not to be represented in the Antarctic. The Haptophyceae are also apparently absent in the coldest waters. Although blooms of blue-green algae have been reported from sub-arctic freshwater lakes (Billaud, 1967) the common representatives of the blue-green algae in the marine plankton, *Trichodesmium* spp., are only abundant in the tropics. The detailed biogeography of phytoplankton cannot be considered here and the reader is referred to Smayda (1958), Braarud (1962), and Johnson and Brinton (1963) for further details. However, as examples of species characteristic of particular regions one may mention the polar species *Thalassiosira antarctica* and *T. hyalina,* of which the former is confined within the Antarctic Convergence, whereas the latter has an eastern Arctic tendency; the tropical species *Planktoniella sol,* which can be used as an indicator of the Gulf Stream near its entrance into the Norwegian Sea; and *Thalassionema nitzschioides,* which is cosmopolitan. The distributions of such species show distinct relationships to salinity as well as to temperature (Smayda, 1958).

Buoyancy is a most important factor in the waxing and waning of phytoplankton populations and must be dealt with at some length here although it has been considered by Lund (1959), Hutchinson (1967), and Smayda (1970). Obviously, in order to photosynthesize, a phyto-plankton cell must remain near the water surface, and this implies a minimal sinking rate or else motility. Positive buoyancy is disadvan-tageous because intensities of visible and ultraviolet radiation at the water surface are often inhibitory and because concentration of cells at the surface limits the volume of water from which they can absorb nutrients. On the other hand, sinking is not without its advantages since it is a means of achieving relative motion between a cell and the water in its immediate vicinity. Munk and Riley (1952) pointed out that transfer of dissolved substances between environment and cell will be faster if the concentration gradient is kept steep in this way, *i.e.,* if nutrients are continually renewed in the impoverished zone adjacent to the cell. Swimming or sinking will provide such "forced convection." Munk and Riley derived formulae for the rate of nutrient absorption by various algae approximating to shapes of spheres, discs, cylinders, and plates, and showed that division rates in diatoms are related to their calculated sinking rates.

Most plankton algae belonging to the Prasinophyceae, Chryso-phyceae, Haptophyceae, Cryptophyceae, and Dinophyceae possess fla-gella by means of which they can swim, and it seems likely that this ability to obtain forced convection is of biological advantage to them. It is perhaps more than a coincidence that nearly all the phytoplankton species requiring vitamin B_{12} , a particularly scarce nutrient, are motile (Hutchinson, 1967). Dinoflagellates studied by Eppley *et al.* (1968) were found to swim at rates of 1 to 2 meters per hour and some showed a clear-cut migration to the surface during the daylight hours and into deeper layers at 5 to 16 m at night, both in the sea and in laboratory cultures in a 10 m water column (Fig. 23). Such movement, besides providing forced convection, enables excursions into water layers which may be more nutrient rich than those near the surface. It has been shown that some populations of dinoflagellates, such as are found in red tides, could only have obtained the amount of nitrogen they contain by "sweeping out" the water-column in this way (Eppley *et al.,* 1968, 1969). Freshwater dinoflagellates have been observed to show a similar diurnal vertical migration (Berman and Rodhe, 1971).

Nevertheless, two of the most successful plankton groups, the dia-toms and the blue-green algae, are not able actively to swim. These, at first sight, would appear to be faced with the insoluble problem of

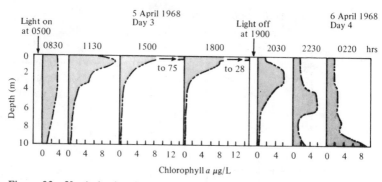

Figure 23. Vertical migration of the marine dinoflagellate *Cachonina niei* in a deep tank as demonstrated by depth profiles of chlorophyll *a* at different times of the day. From R. W. Eppley, O. Holm-Hansen, and J. D. H. Strickland, Some observations on the vertical migration of dinoflagellates, *J. Phycol.* (1968), *4:* 333–340, fig. 7 (Phycological Society of America Inc.).

remaining near the surface to obtain light for photosynthesis and, at the same time, sinking to achieve forced convection and so absorb sufficient nutrients to maintain a high growth rate. That they are able to reconcile these requirements appears to be largely due to the wind-induced circulation which occurs in the surface layers of the water. Wind blowing across a water surface causes it to circulate in long "cells" parallel or at a slight angle to the wind direction (Fig. 24). The occurrence of this phenomenon, which is called Langmuir circulation after its discoverer, is indicated by the parallel slicks, spaced from a few centimeters to 50 or more meters apart and marking the lines of downwelling, which can nearly always be seen on all but the calmest and roughest waters (Plate 2). Since the upwelling speeds (Fig. 24), even at low wind velocities, greatly exceed the sinking speeds of phytoplankton, this means that some of a phytoplankton population will be retained in circulation in the "cell," only those cells in the region of most rapid downwelling having their descent into the depths accelerated (Fig. 25). In this way the population may be able to multiply and maintain itself in the light near the surface but, at the same time, with the cells sinking relative to the water in their vicinity and so achieving good conditions for growth. This will happen at the sacrifice of a proportion of the population as illustrated in Figure 30 (p. 90).

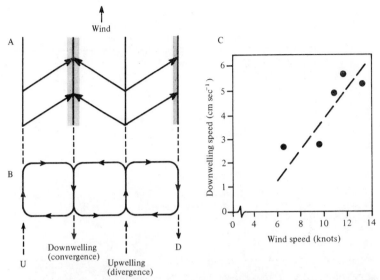

Figure 24.   Langmuir circulation. A, surface pattern illustrating wind-row or slick formation (stippled area) at lines of convergence; vectors represent surface water movement. B, vertical cross-section of the circulation "cells." C, the relationship between the downwelling speeds at convergences and the wind speed. Data of Woodcock (1950) and Sutcliffe, Baylor, and Menzel (1963) from T. J. Smayda, The suspension and sinking of phytoplankton in the sea, *Oceanogr. & mar. Biol. Rev.* (1970), *8*:353–414, fig. 6 (George Allen and Unwin, London).

To exploit this situation to best advantage it is necessary that the sinking rate should have an optimum value in relation to that of speed of upwelling. It is perhaps unlikely that any alga could adjust its sinking rate to match the hour to hour variation in Langmuir circulation but one might expect some correspondence between the sinking rate of a species and the average conditions under which it is usually abundant. There are many features which might be concerned in regulation of sinking rate and to consider them logically one may take Stokes's law as a starting point. This states that for a sphere of radius, r,

$$\hat{v} = \tfrac{2}{9} g r^2 \, (\rho' - \rho) \, \eta^{-1}$$

where $\hat{v}$ is the terminal velocity of sinking; g, the gravitational constant; ρ', the density of the sphere, and ρ, that of the liquid; and η, the viscosity of the liquid. Sinking rates of plankton organisms calculated

Plate 2. Wind-rows in Port William, Falkland Islands (51° 40′ S, 57° 47′ W), March 1974. The wind-rows, which were about 1 m apart, indicate lines of downwelling associated with wind-induced Langmuir circulation. Photograph by G. E. Fogg.

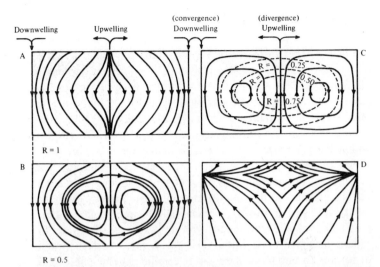

Figure 25. The trajectories of phytoplankton cells as affected by Langmuir circulation under different conditions. To the right of the upwelling (convergence) area the water circulates in a clockwise direction and to the left in a counterclockwise direction. A, the situation when the sinking speed of the phytoplankton cells is just greater than the maximum speed of upwelling (R = 1). B, trajectories of phytoplankton cells the sinking speed of which is half of the maximum speed of upwelling (R = 0.5); in the central area of closed trajectories the phytoplankton cells swirl around and are retained, outside this area they sink out of the circulation "cell." C, here the solid lines represent the streamlines of Langmuir circulation; the dashed lines are the boundaries of the regions of retention for phytoplankton cells sinking at various speeds relative to that of downwelling speed (R values). D, trajectories of a motile species tending to maintain itself just below the surface; the cells collect where the trajectories converge on the right and on the left. From T. J. Smayda, The suspension and sinking of phytoplankton in the sea, *Oceanogr. & mar. Biol. Rev.* (1970), *8:* 353–414, fig. 7 (George Allen and Unwin, London).

by this equation show reasonable agreement with observed values (Hutchinson, 1967).

First, it should be noted that Stokes's law only applies to small spherical objects. Departure from the spherical shape will increase the surface/volume ratio, thereby increasing the friction with the liquid and so causing the object to sink more slowly. This can be expressed in terms of a coefficient of form resistance, which is the ratio of the terminal velocity of a sphere of the same volume to that of the actual object. For a disc one hundred times as wide as it is thick this coefficient varies from 2.9 to 4.3 according to whether it falls edge on or broadside on. The striking elaboration of form encountered among phytoplankton species (see the frontispiece) is probably of biological value in decreasing rate of sinking (it will also, of course, increase the relative rate of nutrient absorption [p. 21] and, possibly, by presenting a more awkward mouthful, discourage grazing animals). The fact that, within the size range of net plankton, a cylinder falls more slowly than a disc or sphere of the same volume, if it is sufficiently long, may provide an explanation for the frequency of long filamentous and acicular forms in the phytoplankton (Hutchinson, 1967; Smayda, 1970). It has been shown experimentally (Conway and Trainor, 1972) that colonies of *Scenedesmus* with spines or bristles have a slower sinking rate than those without. However, on the basis that elaboration of form is an important means of controlling sinking rate, it is difficult to explain the observation that many species are enveloped in mucilage to the extent of rendering them effectively spherical, as may easily be demonstrated by mounting them in negative stain, which presumably offsets the advantages which elaboration of form confers.

According to Stokes's law sinking rate increases with increasing cell size, and this has been confirmed by observations on diatoms (Eppley *et al.,* 1967). If there is a tendency towards aggregation of cells, that is, formation of larger particles, then sinking rate becomes a function of cell concentration. Eppley *et al.* (1967) found this particularly evident with *Thalassiosira fluviatilis,* which releases long thin strands of chitan in which the chains of cells become entangled.

Apart from decreasing its linear dimensions, the most effective way of decreasing sinking speed open to an alga is to decrease its specific gravity relative to that of the water. This may be accomplished by accumulation of some material of low specific gravity, the possibilities being as follows:

1. *Fat accumulation.* It is generally supposed that many plankton species are able to float because their mean specific gravity is lowered by accumulation of fat. Undoubtedly, a form such as *Botryococcus braunii,* which contains lipoids up to 30 or 40 per cent of its dry weight, floats for this reason, but the evidence that fat accumulation

enables diatoms to float is rather poor. Diatoms with conspicuous fat droplets in their protoplasts often have thickened silica walls, so that the cells are heavy and sink. Furthermore, laboratory experiments (p. 55) suggest that fat accumulation is related to cell breakdown, and hence, that cells containing large amounts of fat are perhaps not viable.

2. *Mucilage production.* This is common among phytoplankton species and reduces the sinking speed provided that the difference in density between cell and mucilage is at least twice the difference in density between mucilage and water (for the mathematical demonstration of this see Hutchinson, 1967).

3. *Selective accumulation of ions.* Gross and Zeuthen (1948) put forward evidence suggesting that the specific gravity of the marine diatom *Ditylum brightwellii* is maintained at a low value by preferential absorption of monovalent as against divalent ions—a solution containing only monovalent ions such as Na^+ and Cl^- having a lower specific gravity than an isotonic solution containing divalent ions as well. Beklemishev *et al.* (1961) have shown by direct analysis that the cell sap of the giant diatom *Ethmodiscus rex* does actually contain reduced concentrations of divalent ions, especially magnesium, as compared with seawater. This situation would need to be maintained by active metabolism, and various workers have observed that living cells of diatoms have slower sinking rates than corresponding dead ones. Living cells of *Coscinodiscus wailesii*, for example, have been observed to sink 5 to 9 m per day whereas dead cells sink 30 m per day (see Smayda, 1970). Eppley *et al.* (1967) found that sinking rate was inversely related to growth rate in cultures of *Thalassiosira fluviatilis* (Fig. 26). Similar behavior seems to be shown also by the green alga *Scenedesmus* (Conway and Trainor, 1972). Since it is presumably the vacuole of the cell in which the selective accumulation of ions occurs, this mechanism would be expected to be most effective in large species with highly vacuolated cells (Braarud, 1962). Although the suggestion of Gross and Zeuthen is plausible for marine phytoplankton, it does not seem possible that it can apply in fresh water, the density of which is only of the order of 0.003 per cent greater than that of pure water. Observations on the freshwater species *Asterionella formosa* under a variety of conditions show that the cells are always heavier than water and sink if there is no turbulence (Lund, 1959).

Smayda (1970) concluded that notwithstanding the attractiveness of

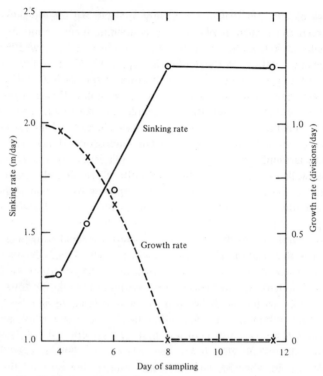

Figure 26. Sinking rate in relation to growth rate in a culture of the marine diatom *Thalassiosira fluviatilis* approaching the stationary phase of growth. The sinking rates shown have been corrected to allow for the aggregation of cells which occurs with increasing cell concentration. From R. W. Eppley, R. W. Holmes, and J. D. H. Strickland, Sinking rates of marine phytoplankton measured with a fluorometer, *J. exp. mar. Biol. Ecol.* (1967), *1*:191–208, fig. 7 (North-Holland Publishing Co., Amsterdam).

the selective ion accumulation theory it remains to be confirmed except perhaps in the anomalous dinoflagellate *Noctiluca*.

4. *Gas-vacuoles*. Many blue-green algae have the peculiar characteristic, found among no other organisms except certain bacteria, of having spaces containing gas within their protoplasts.* These gas-vacuoles undoubtedly lower the specific gravity of the cells, often to

---

*A few protozoa, such as *Arcella* spp., may contain gas bubbles but these are of an entirely different nature from the gas-vacuoles of prokaryotes.

such an extent that they are positively buoyant and some planktonic blue-green algae, such as *Microcystis,* containing them, commonly accumulate in a thick layer at the water surface in calm weather, a phenomenon known as "water-bloom" (p. 108). Most of the water-bloom-forming blue-green algae are freshwater species but *Trichodesmium* spp. are marine forms that have gas-vacuoles. If such algae are subject to sudden pressure the gas-vacuoles are collapsed, the optical properties of the cells undergo an immediate change, and the cells sink instead of floating (Plate 3). Electron microscopy shows that gas-vacuoles are built up of regular arrays of cylindrical gas-vesicles (Plate 3). The walls of these vesicles, which consist of protein only, are freely permeable to gases, and the vesicle thus appears to contain a space, created within the protoplast, which fills with gas in equilibrium with that in solution in the cell (Walsby, 1972).

Gas-vesicles provide a means for comparatively rapid adjustment of buoyancy. The quantity of gas-vesicles which a cell contains depends on the rate at which they are formed in relation to the growth of the alga. Thus in dim light, such as prevails at the bottom of the photic zone, the rate of cell division is low so that vesicles accumulate and the cells become more buoyant. On the other hand, near the surface, under conditions favoring active photosynthesis and growth, gas vesicles will be "diluted out" by growth and the cells become less buoyant. This provides a poising mechanism which will result in the accumulation of the population at some intermediate level (Fogg *et al.,* 1973). It is, in fact, commonly observed that blue-green algae have well-defined population maxima at particular depths in the water-column (Fig. 27; Zimmermann, 1969; Brook *et al.,* 1971; Walsby and Klemer, 1974). In addition, gas-vesicles may be collapsed by increase in the turgor pressure of the cell, to which, of course, they are subjected as well as to the external hydrostatic pressure (Walsby, 1971). Dinsdale and Walsby (1972) have shown very elegantly that gas-vesicles are collapsed by the increase in turgor pressure due to the accumulation of osmotically active photosynthate, and Reynolds (1973) has shown that in a natural population of *Microcystis,* the gas-vacuole content of cells and their buoyancy are in inverse relation to their turgor pressure (Fig. 28). If active growth occurs the products of photosynthesis will be used for synthesis of cell materials of low osmotic activity so that active collapsing of gas-vesicles would not be expected in the presence of an

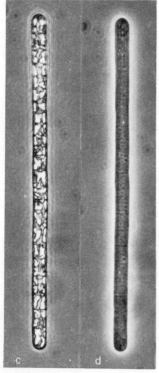

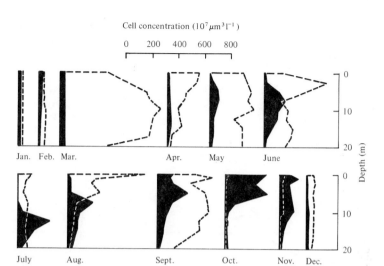

Figure 27. Vertical distribution of the blue-green alga *Oscillatoria rubescens* (black) and other phytoplankton (dotted lines) in the Vierwaldstättersee (Lake Lucerne), Switzerland, in 1966. Data of Zimmermann (1969) from G. E. Fogg, W. D. P. Stewart, P. Fay, and A. E. Walsby, *The Blue-green Algae* (1973), fig. 12.5 (Academic Press, London).

Plate 3. The gas-vacuoles of blue-green algae. (a) and (b) The "hammer, cork and bottle" experiment, showing the results of collapsing gas-vacuoles in *Microcystis aeruginosa;* (a) the decrease in turbidity after striking the cork in the right-hand bottle, thus destroying the gas-vacuoles by increased pressure (the bottle on the left is untreated); (b) the loss of buoyancy ⸱pparent in the treated sample after allowing it to stand for two hours. After A. E. Walsby, Structure and function of gas-vacuoles, *Bact. Rev.* (1972), *36*:1–32, fig. 1 (American Society for Microbiology). (c) Filament of *Oscillatoria agardhii* var. *isothrix* with intact gas-vacuoles (X 800); (d) the same filament after the gas-vacuoles have been collapsed by application of pressure. After A. E. Walsby and A. R. Klemer, The role of gas vacuoles in the microstratification of a population of *Oscillatoria agardhii* var. *isothrix* in Deming Lake, Minnesota, *Arch. Hydrobiol.* (1974), *74*:375–392 (Schweizerbart'sche Verlagsbuchhandlung, Stuttgart). (e) Cell of *Anabaena flos-aquae* showing the cylindrical gas-vesicles, which comprise the gas-vacuoles, freeze-fractured lengthwise and in cross-section (X 41000). Micrograph by D. Branton and A. E. Walsby from Gas vacuoles, in *The Biology of the Blue-green Algae,* ed. N. G. Carr and B. A. Whitton, pp. 340–352, fig. 16.2 (Blackwell Scientific Publications, Oxford).

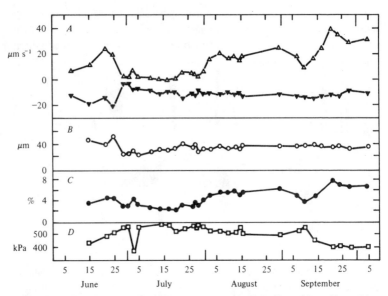

Figure 28.   The blue-green alga *Microcystis aeruginosa* in Crose Mere, Shropshire, England, in the summer of 1971; seasonal variations in (A) flotation and sinking rates in fresh (▲) and non-gas-vacuolate (△) suspensions; (B) mean colony radius; (C) relative gas-vacuole volume, and (D) mean cell turgor pressure. From C. S. Reynolds, Growth and buoyancy of *Microcystis aeruginosa* Kütz, emend. Elenkin in a shallow eutrophic lake, *Proc. R. Soc., B* (1973), *184*:29–50, fig. 6 (The Royal Society of London).

ample nutrient supply. In this case the population would be expected to maintain a position higher in the water-column than if nutrients were limiting. Observations such as those of Zimmerman (1969) on the seasonal migrations of population maxima of *Oscillatoria rubescens* fit in with this hypothesis (Fogg *et al.*, 1973).

Alterations in buoyancy by changes in gas-vacuole content are evidently sufficiently rapid to result in a diurnal vertical migration, as has been observed with *Microcystis* in Lake George, Uganda, by Ganf (1969). Thus, apart from serving to locate a blue-green alga at a depth in the water-column most favorable for growth, the gas-vacuole mechanism also provides for forced convection and perhaps for descent to nutrient-rich water towards the bottom of the photic zone. Other ecological aspects of gas-vacuole-containing blue-green algae are considered in a later chapter (p. 109).

It may be noted that cells may also decrease their specific gravity by decreasing their load of heavy components. Members of the Coccolithophoridaceae may thus reduce sinking rate by shedding coccoliths, which being composed of calcium carbonate are relatively heavy. Coccolith-bearing cells of *Coccolithus huxleyi* have been shown to sink five times as fast as otherwise similar naked cells (Eppley *et al.*, 1967).

Finally, returning to Stokes's law, it is conceivable that some control of sinking rate might be achieved by small cells altering the arrangement of water molecules in their vicinity, and thus affecting structural viscosity, through the chemical and electrostatic properties of their cell surfaces. However, no direct evidence of this appears to have been reported.

Increase of Phytoplankton in Temperate Waters in the Spring

The spring maximum in temperate waters is perhaps the best known feature of phytoplankton development, and it is convenient to discuss the effects of most environmental factors in relation to this phenomenon. In many respects it resembles growth in a culture of limited volume. Thus it starts from a small initial number of cells in a medium comparatively rich in nutrients. It may follow an approximately exponential course for several weeks, and often a single species predominates. The final crop is sometimes roughly proportional to the initial amount of a limiting nutrient, as, for example, with *Asterionella formosa* in Windermere (p. 93). From a comprehensive series of observations which included measurements of chlorophyll, photosynthesis, and respiration, during the development of a mixed phytoplankton bloom in St. Margaret's Bay, Nova Scotia, in March and April, Platt and Subba Rao (1970) concluded that it passed through exponential, stationary, and senescent phases analogous to those in a batch culture. The changes in metabolic pattern recorded seem similar to those described in Chapter IV except that no evidence of accumulation of fat was found in the senescent stage.

In the autumn and early winter nutrient concentrations in temperate seas or lakes build up as a result of release through bacterial activity continuing at a high rate relative to algal uptake and mixing in of nutrient-rich deep water. For the renewal of phytoplankton growth the first prerequisite is a suitable inoculum. The two principal possibilities are that a species is always present in the water, or that a species is absent from the water for substantial periods during which its resting stages are present in bottom deposits, from whence the open waters can be repopulated when conditions become favorable again. These types were distinguished by Haeckel (1890) as *holoplanktonic* and *meroplanktonic*, respectively. Good examples of both kinds are known. Lund (1949) found no evidence that appreciable numbers of *Asterionella formosa* in Windermere are derived from sheltered bays or enter by way of the inflowing streams. Resting spores of this species have never been observed. Evidently, live cells are always present in open water, and the observed increases are clearly due to multiplication in the water itself. Among marine phytoplankton *Halosphaera viridis* and *Coccolithus huxleyi* (Braarud, 1962) are examples of holoplanktonic species. It may be noted that a holoplanktonic species does not necessarily persist in a given area of sea throughout the year. *Coccolithus huxleyi*, for example, is presumed by Braarud (1962) to die off in northern and southern waters in the winter and to be re-introduced each season from warmer parts.

Melosira italica provides a clear example of a meroplanktonic freshwater alga. It produces resting stages which lie dormant in the bottom mud until conditions are favorable for their development in the water (Lund, 1954). *Cricosphaera carterae* is an extreme case of a marine meroplanktonic species, having in its life cycle a sedentary phase which may give rise to a motile phase capable of forming dense planktonic growths. *Skeletonema costatum* is a widely distributed meroplanktonic species of coastal waters which does not have a specially differentiated resting stage (Braarud, 1962). Oceanic species are characteristically holoplanktonic, whereas meroplanktonic species must be coastal in origin, although they may occur in oceanic waters if these are inoculated with coastal waters. Thus Paasche (1960*b*) attributed the abundant occurrence of the meroplanktonic species *Chaetoceros debilis* in the Norwegian Sea in 1954 to an inoculation from the Faroe region into the Atlantic water masses entering this area.

By mid-winter the stage has been set for algal growth, but physical conditions are limiting. Lund *et al.* (1963) found that *Asterionella* grew as vigorously in water taken from Windermere after the middle of November as it did in culture solution. Light and temperature in winter are low, but the intensity and duration of light are sufficient to support some algal growth, and abundant development of phytoplankton can frequently occur at temperatures approaching zero; development of plankton under ice is common. Nevertheless, phytoplankton numbers are generally low in winter. This is presumably because the rate of loss of whole cells, or of cell material by respiration, exceeds or approximately balances the rate of addition by growth. Increase does not become perceptible until early spring, when it begins abruptly, often showing, as with diatoms in the English lakes, a remarkably similar timing year after year.

The reasons for this abrupt beginning of population increase seem to be fairly clear. Gran and Braarud (1935) pointed out that net growth cannot occur if mixing takes place at such a rate and to such a depth that phytoplankton is carried out of the photic zone faster than it can multiply. The point at which growth exceeds depletion will be determined by both the prevailing light intensities and the vertical coefficient of eddy diffusion.* The relation between the critical depth of mixing and the onset of the spring increase has been considered more precisely by Sverdrup (1953; see also Ryther, 1963). In temperate seas there is good correlation between the beginning of phytoplankton increase in the spring and the time at which the depth of the surface mixed layer becomes less than the critical depth calculated from the amount of incident radiation and the extinction coefficient of the water. Where the water-column is relatively stable during the winter, as in the Kattegat, there may be a high rate of phytoplankton production at this season (Steemann Nielsen, 1964b). A generally similar situation in relation to mixing no doubt occurs in lakes, but other factors in addition to turbulence may operate to keep the rate of depletion of the population greater than its rate of growth. Thus Lund et al. (1963) consider that Asterionella populations in Windermere are depleted

*When heat or a material is transported by turbulence, the amount, S, passing per second through a surface of 1 cm^2 at right angles to the gradient is given by the expression, $S = A\ s'$, where s' is the gradient per cm and A is the coefficient of eddy diffusion. A, of course, is not a constant peculiar to a given substance, in this case water, but depends on the turbulence of the medium.

mainly by loss in the outflow, with losses from sedimentation onto the deposits, ingestion by animals, and parasitism by fungi being slight. The wetter and colder the winter, the smaller is the size of the population at the end of it.

There is no direct evidence of any event in the sequence resembling the lag phase in culture—having regard to the extremely low density of the populations concerned it is to be expected that such evidence would be difficult to obtain—but nevertheless it may be argued that something of this sort is involved. As we have already seen (p. 14), the lag phase is most pronounced in populations of low density. If a cell of a plankton alga requires a certain minimum concentration of some external metabolite in its vicinity before cell division can begin, then a dependence of increase in numbers on light intensity and decrease in turbulence, such as is observed, would be expected. The production of extracellular products is dependent on photosynthesis and thus on light intensity, and if the turbulent mixing extends to the vicinity of the cells, it will delay the establishment of sufficiently high concentrations of these substances around them. Glycolic acid is one external metabolite which may be important in this way, and there are presumably others. The physiology and biochemistry of the liberation of glycolate by algae has now been studied extensively, and there is evidence that the process occurs in natural phytoplankton populations both in freshwater and the sea (Fogg, 1971, in the press a). As we have seen (p. 16), the presence of glycolic acid reduces the lag phase in cultures. Glycolic acid has been detected in lake waters in concentrations up to 0.06 mg per liter (Fogg et $al.$, 1969). It occurs in the sea in similar concentrations which fluctuate in parallel with the amount of phytoplankton, with maxima in early summer and autumn and zero concentration in mid-winter (Al-Hasan et $al.$, in the press).

In a study of the spring outburst, which was dominated by the diatom $Stephanodiscus$ $hantzchii$, in Lake Erken, Sweden, Pechlaner (1970) found that the initiation of the increase was related to the increase in the availability of radiation. However, a lag in response seemed to be caused by the need for adaptation to the higher light intensities. The efficiency of photosynthesis (carbon fixed per unit of chlorophyll per unit of radiant energy per day) increased over the first 3 or 4 weeks of the outburst. Since there was scarcely any change in water temperature during this time, this increase could not have been due to temperature-dependent speeding up of enzyme-catalyzed reactions.

There is thus some indication of a lag-phase resembling that in culture, and it may be noted here that Johnston (1963*b*) has commented that from biological assay results "poor quality has been the general rule for sea waters before the spring bloom commences. Presumably some modification of sea water must take place when light and stability become suitable, before the phytoplankton bloom is in full swing."

Increase during the spring outburst is sometimes of only one dominant species and usually follows an approximately exponential course (Fig. 29). Considering the errors of sampling and counting mentioned in the previous chapter and the fluctuations of light and temperature which occur in the natural habitat, it is not surprising that the points do not always lie along a straight line when numbers are plotted logarithmically.

It is important to note that it is not strictly correct to speak of "growth rate" in connection with natural populations of phytoplankton; "rate of increase" is better, since the same population is not necessarily being sampled all the time, and since depletion by removal or death of cells is going on concurrently with multiplication.

It will be noticed from Figure 29 that the relative rates of increase of *Asterionella* under lake conditions correspond to doubling times of 5 to 7 days. This is much longer than the time under laboratory conditions, viz., 9.6 hours at 20° C (see Table 2). This difference is mainly attributable to the fact that the bulk of the lake population is light and temperature limited, although loss of cells from the photic zone accounts for part of it (Fig. 30).

Since the depth of the photic zone is often 10 meters or less, and never more than 100 meters, the greater part of the water is usually unavailable for phototrophic growth. If chemotrophic growth of phytoplankton occurs, even at a slow rate, it might be important in contributing to the total crop, simply because, being independent of light, it could occur throughout the water column. It will be recalled from Chapter II that although many algal species are capable of growth in the dark if given a suitable organic substrate, there seems to be little evidence from laboratory cultures of truly planktonic species being able

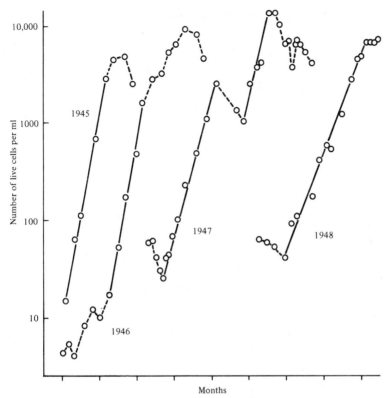

Figure 29. Numbers of live cells of *Asterionella formosa* per ml in Esthwaite Water 1945–48, plotted logarithmically against time. Periods of approximately exponential increase are denoted by the solid straight lines. Data of Lund, 1950.

to do this. Although both fresh waters and sea waters normally contain relatively high concentrations of dissolved organic substances, the concentrations of readily assimilable substances such as sugars and amino acids are generally low (Duursma, 1965). Glycolic acid may, perhaps, be a readily assimilable substrate available in quantity, and it must be remembered that a substance in low concentration may nevertheless have a high turnover rate and be an important energy source for growth.

Rodhe (1955) found indications of growth of nannoplankton in a sub-arctic lake during the winter under circumstances in which virtually

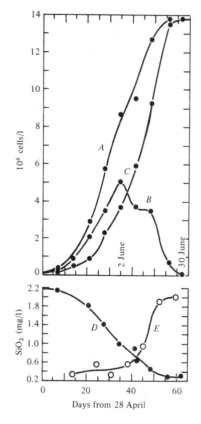

Figure 30. Production and loss of *Asterionella formosa* in Windermere, North Basin, April 28 to June 30, 1947. The epilimnion is assumed to occupy the top 8 m, and the curves represent: *A*, the cumulative total of cell production computed from silicate uptake: *B*, the mean concentration of cells in the epilimnion (standing crop); *C*, the cumulative loss of cells from the epilimnion (*A* minus *B*); *D*, epilimnetic silicate concentration; *E*, relative rates of loss of cells from unit epilimnetic population (arbitrary scale). From J. W. G. Lund, F. J. H. Mackereth, and C. H. Mortimer, Changes in depth and time of certain chemical and physical conditions and of the standing crop of *Asterionella formosa* Hass. in the north basin of Windermere in 1947, *Phil. Trans. R. Soc., B* (1963), *246*:275, fig. 9.

no light could have been available. He supposed that chemotrophic growth was taking place at the expense of dissolved organic matter produced by photosynthesis during the summer. Bernard (1963) has amassed a great deal of evidence which shows that nannoplankton algae are often as numerous in samples taken from considerable depths in the sea as they are in the photic layer. It seems that chemotrophic growth must be occurring under these circumstances. Parsons and Strickland (1962) showed by use of ^{14}C as a tracer that there is definite uptake of organic substrates by natural phytoplankton populations, but it cannot be said from their results whether this is due to the algae themselves or to bacteria associated with them. Using a development of this method based on enzyme kinetics Hobbie and Wright (1965) obtained results which they took as showing that, in natural populations, phytoplankton algae compete poorly with bacteria for organic substrates. Other workers (Bunt, 1969; Pant, 1973), however, have experienced difficulty with this method and conclude that results obtained by its means must be regarded with caution.

The detailed work of Lund *et al.* (1963) showed no growth of *Asterionella* in the aphotic layers of Windermere. On the whole, however, the evidence for chemotrophic growth on the part of some phytoplankton species, at least, seems strong, though certainly far from conclusive. Phototrophic assimilation of organic substances, for which there is good evidence from laboratory experiments (p. 28), is also a possibility that must be borne in mind.

The factors which may operate to bring to an end this period of rapid increase in the spring are various and not necessarily the same as those which operate in cultures of limited volume. It is particularly important to remember in this connection that the population densities in cultures and in natural waters are usually of entirely different orders of magnitude. An ordinary laboratory culture of *Chlorella* may contain 10^6 cells per mm^3 when growth is complete, whereas a dense *Asterionella* bloom in Windermere has only 10 cells or so per mm^3. Even allowing for the fact that *Asterionella* cells have about 10 times the volume of *Chlorella* cells, there is an enormous difference here, and one might expect the factors controlling growth to differ accordingly.

Light Intensity

As the concentration of cells in the water increases, light penetration decreases, so that the phytoplankton becomes self-shading. Inverse correlations between plankton density and light penetration have been described frequently, but it is not always easy to distinguish between effects due to the plankton itself and, for example, detritus brought in by floods. Reduction of light penetration will have the same effect on growth, whatever its cause, but it is clear that chance occurrences such as floods are not usually responsible for cessation of growth, and that, if reduction of light is concerned at all, self-shading is the important factor. Talling (1960*b*) has reported a particularly clear example of self-shading by *Asterionella* in Windermere. During 1959 a well-developed peak in *Asterionella* occurred during a period of bright, calm

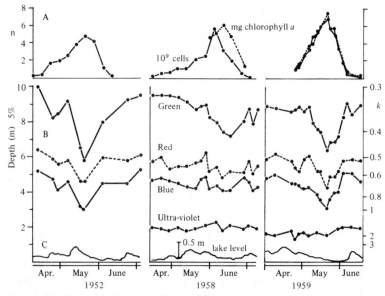

Figure 31. Seasonal changes in Windermere, during parts of three years, of (A) numbers of cells of *Asterionella formosa* (in 10^9 cells per m³; depth range 0-5 m in 1952 and 1958, 0–10 m in 1959), and the chlorophyll *a* content in mg/m³ (depth range 0-10 m); (B) the penetration of light in 3 or 4 spectral regions expressed as the depth interval in which a reduction to 5 per cent occurs, with the corresponding scale of the vertical extinction coefficient (k) also indicated; (C) lake level. From J. F. Talling, Self-shading effects in natural populations of a planktonic diatom, *Wett. Leben* (1960), *12*:239, fig. 1.

weather with no flooding, when other plankton species were scarce. Under these conditions there was a close correlation between *Asterionella* numbers and light extinction (Fig. 31). Talling concluded that only rarely does phytoplankton reach a density sufficient to have appreciable effects in reducing light penetration. In the present context it is to be noted that the reduction in light penetration, through self-shading, while large enough to have some effect on photosynthesis and hence on growth, is not drastic. It could only account for a decrease in growth rate and not for a complete cessation in increase, such as is actually observed.

In passing it should be recorded that adaptation of phytoplankton to light intensity, similar to that found in cultures (p. 24), occurs during periods of stability of the water column (Ryther and Menzel, 1959; Steemann Nielsen and Hansen, 1959).

Mineral Nutrients

Deficiency of a mineral nutrient may be expected to be one of the most important factors causing cessation of spring growth, but clear instances of this are few. One has been provided by Lund (1950) for *Asterionella* in the English Lake District. Detailed records of the numbers of this diatom and of the levels of various environmental factors during the spring period have been accumulated over many years. Factors such as light intensity, temperature, grazing by zooplankton, and parasitism did not seem to be responsible for the cessation of growth, and, in particular, there was no correlation of this with a fall in concentration of either nitrate-nitrogen or phosphate to any definite level. Furthermore, the concentrations of these nutrients left in the water after the decline in population were sufficient to support appreciable further growth. Similarly, amounts of inorganic carbon and calcium seem to have been sufficient to provide for much larger populations than were observed. On the other hand, the decrease in numbers of *Asterionella* has coincided over a period of many years (Lund, 1950; Lund *et al.*, 1963) with a drop in dissolved silica concentration below 0.5 mg/liter (see Fig. 32). This agrees with an observation of Pearsall (1932) that diatoms cannot multiply appreciably if the concentration of silica is below this level. Jørgensen (1957) found the growth of various freshwater diatom species both in culture and in lakes to be limited only at silica concentrations somewhat lower than this. Observations on marine species do not seem to justify any definite conclusions, and *Skeletonema costatum* appears to be able to grow, producing a very thin siliceous wall, when the silica supply is greatly reduced (Braarud, 1962).

It is a matter of some difficulty to decide in any given instance whether a particular nutrient is limiting phytoplankton growth or not. The concentration found in the water represents the balance between consumption and supply, and a low concentration is not necessarily an indication that a nutrient is in short supply. As we have already seen (p. 40), the concentration at which a given nutrient becomes limiting varies according to the level of other factors, since cells may accumulate substantial reserves of certain nutrients, and there are considerable

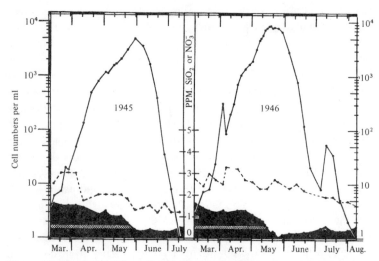

Figure 32. Numbers of live cells of *Asterionella formosa* per ml in Windermere, North Basin, during parts of 1945 and 1946, plotted logarithmically against time (solid line); interrupted line, nitrate nitrogen in parts per million, values multiplied by 10 for convenience in representation; solid black, dissolved silica in parts per million with the 0.5 parts per million level cross-hatched. After J. W. G. Lund, Studies on *Asterionella formosa* Hass. II. Nutrient depletion and the spring maximum, *J. Ecol.* (1950), *38*:6, fig. 3 (Blackwell Scientific Publications Ltd., Oxford, Eng.).

discrepancies between observations in culture and in natural waters. The rather good agreement between laboratory and field observations on silica limitation of diatoms just described is perhaps exceptional and related to the special metabolic interrelations of this element and its low turnover rate in the water but even this agreement is dependent to a certain extent on the level of other factors. Hughes and Lund (1962) found that the addition of small amounts of phosphate to Windermere water permitted the growth of so large a crop of *Asterionella* that all the silica present was incorporated into the cells. The regular observed relationship between the spring maximum of *Asterionella* in Windermere and silica concentration thus seems to be contingent on the concentration of phosphate in the lake water and the rate of growth of the diatom as determined by the prevailing temperature and light conditions.

Gerloff and Skoog (1954, 1957) proposed that the cell contents of elements such as nitrogen and phosphorus may be used as a measure of their availability in the water. As pointed out in Chapter IV, reserves of both these elements may be accumulated in cells so that division may take place in the absence of a further supply until certain limiting intracellular concentrations are reached. Gerloff and Skoog studied *Microcystis aeruginosa* in particular and determined in laboratory experiments the cell concentrations of phosphorus and nitrogen below which cell division would not occur. The results given in Figure 33, for example, show that yield diminishes when cell nitrogen falls below 4 per cent on a dry weight basis. Gerloff and Skoog argued that if analysis of *M. aeruginosa* from a lake shows a value higher than this, then nitrogen cannot be limiting for growth. The method can be refined by making allowance for the variable production of mucilage by this species in the estimation of dry weight. On this basis Gerloff and Skoog concluded that nitrogen, but not phosphorus, limits growth of M. *aeruginosa* in the lakes of southern Wisconsin. Soeder *et al.* (1971) list values for the minimum phosphorus content, which they believe to be reasonably constant and species specific, of various freshwater algae. Comparison of phosphorus contents with these minimum values should

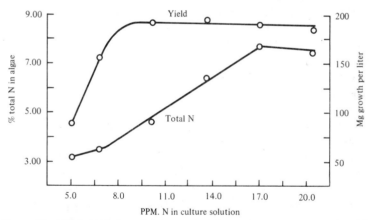

Figure 33. Total yield (mg dry wt per liter) and total nitrogen content as a percentage of dry weight of *Microcystis aeruginosa* grown for 14 days in medium containing various levels of nitrate-nitrogen. From G. C. Gerloff, and F. Skoog, Cell contents of nitrogen and phosphorus as a measure of their availability for growth of *Microcystis aeruginosa, Ecology* (1954), *35*:350, fig. 3.

indicate whether phosphorus is limiting or not in a given sample. Gerloff and Fishbeck (1969) have obtained comparable data for calcium, magnesium, and potassium in various freshwater green and blue-green algae. This method seems a useful one, but it is desirable to know how much the limiting cell concentration of an element may change according to the level of other environmental factors. As we have seen (p. 42) variation is to be expected. Fitzgerald and Nelson (1966) have described a method, which seems not to suffer from this uncertainty, in which a 60 min. boiling water extraction separates essential phosphorus compounds in algae from phosphate reserves, which can then be determined in the extract as orthophosphate. The results of such determinations correspond well with the known phosphorus status of various samples and with the phosphatase assay to be described below.

Enzymic activity may be a good index of limitation by certain elements. Phosphorus-deficient algae, for example, show marked increases in phosphatase activity as compared with similar material with ample phosphorus. The activity of alkaline phosphatase can readily be measured at pH 9 using p-nitrophenylphosphate as a substrate; Fitzgerald and Nelson (1966) used this to assess phosphate limitation. Algae which were phosphorus-limited showed 25 times more alkaline phosphatase activity than otherwise similar cells with surplus available phosphorus. With the nitrogen-fixing blue-green algae another method is possible. The uptake of phosphate by phosphorus-deficient material causes a marked stimulation of nitrogen fixation which can be easily detected by the acetylene reduction test. Cells with sufficient phosphate show no such stimulation (Stewart and Alexander, 1971). Stewart *et al.* (1970) used phosphorus-starved *Anabaena flos-aquae* as a bioassay organism in conjunction with this technique in an examination of various Wisconsin lakes and, in agreement with Gerloff and Skoog (1954), found that phosphorus is usually not limiting in these waters.

The rate of ammonium-nitrogen uptake by algae in the dark is 4 to 5 times greater for nitrogen-limited cells as for those with sufficient available nitrogen. Fitzgerald (1968) found this to provide a good basis for determining whether nitrogen was limiting for various freshwater algae. Viner (1973), who applied it in a study of Lake George in Uganda, concluded that nitrogen was the major limiting nutrient in this tropical lake but he obtained results suggesting that in a mixed phytoplankton population different component species may be limited by different nutrients. Nitrate reductase activity is low in nitrogen-limited phytoplankton and its synthesis is induced by nitrate. Eppley *et al.* (1970) found that the nitrate reductase activity of phytoplankton in the Peru current was related to the nitrate concentration in the water. However, since synthesis of the enzyme is repressed by ammonium-nitrogen its activity is not altogether a reliable guide to nitrogen limitation.

Another approach is to supplement samples of water with mineral nutrients, singly or in combination, and observe the effect on the phytoplankton. This is akin to the bioassay technique which is now being widely used to assess the potential fertility of freshwaters. This consists essentially of inoculating one or more test species into filtered water, which may be supplemented with particular nutrients, and incubating under standard conditions of light and temperature after which the amount of algal material is estimated by some appropriate method. Details of suitable techniques have been described by Lund *et al.* (1971). Moss (1969*b*) has provided examples of bioassays on central African waters and discussed some of the difficulties in interpretation. Jensen and Rystad (1973) adapted the dialysis culture technique (see p. 43) for the semi-continuous monitoring of the capacity of seawater to support phytoplankton growth. With three different algae systematic responses to nutrient levels in the seawater external to the cultures were found. The transition point between exponential and linear growth was well correlated with nutrient content and the growth rate in the linear phase was proportional to nitrate concentration. Cell density in the stationary phase was also determined by nitrate concentration.

To minimize errors due to changes which are inevitable in samples in bottles it is best to make measurements for short rather than longer periods. Ryther and Guillard (1959) have done this, using the photosynthetic uptake of ^{14}C from bicarbonate as a measure of algal activity. In water samples taken in the northwestern Atlantic and in the Sargasso Sea they found that rather than nitrogen or phosphorus, silicate and one or more components of their iron-trace metal mixture were limiting photosynthesis. Thomas (1969) found that nitrogen was the element most likely to be limiting in the eastern equatorial Pacific Ocean, however, and it seems probable that this is true for most sea areas (Strickland, 1972). Using the supplementation method, Goldman (1961) found that in Brooks Lake, Alaska, magnesium was the most limiting element in June when there was a peak in phytoplankton with Chrysophyta and Chlorophyta dominating. Addition of phosphate also produced some increase at this time. Later, nitrate also became limiting. In Castle Lake, California, additions of 0.1 ppm or less of molybdenum produced stimulation of photosynthesis in samples taken at various times in the year (Goldman, 1960). This method is not without its pitfalls. Addition of a nutrient element such as potassium, calcium, or magnesium may give better growth by altering the ratio of monovalent to divalent ions (Miller and Fogg, 1957), even though the element itself is not limiting. A salt may contain trace elements as impurities in sufficient amounts to make good a deficiency, and addition of phosphate may result in precipitation and complications due to consequent adsorption of ions.

Carbon Dioxide

All photosynthetic plants can assimilate free carbon dioxide and undissociated carbonic acid without any preliminary induction period. It is still uncertain as to how far the generality of phytoplankton species can utilize, directly, the dissociated forms, particularly bicarbonate (Raven, 1970). The dissociated forms can, of course, give rise to the undissociated forms and thus support photosynthesis, but the physical chemistry of the carbon dioxide system in water is complicated, and it is usually difficult to decide whether dissociated forms are being used directly or indirectly. Perhaps the concentrations of carbon dioxide and/or bicarbonate found in most lake or sea waters are not sufficiently high to be saturating for photosynthesis under high light intensities. Paasche (1964) found that the rate of photosynthesis of *Coccolithus huxleyi* in sea water is carbon dioxide limited at high light intensities. On the other hand, it is not often that the concentration of carbon dioxide and/or bicarbonate in open sea waters is appreciably reduced by photosynthesis. Talling (1960a), for example, found no reduction due to carbon dioxide exhaustion in rate of photosynthesis of *Chaetoceros affinis* in suspension of 19 to 46 cells per mm^3 in natural sea water incubated for periods of 1-3 hours *in situ* in the sea. In dense phytoplankton blooms, however, the rate of carbon dioxide supply may become limiting, especially in fresh water. Wright (1960), for example, found evidence of carbon dioxide limitation of photosynthesis when the density of the standing crop of phytoplankton was high in Canyon

Ferry Reservoir, Montana. Steemann Nielsen (1955) has recorded a particularly clear instance in a Danish lake very rich in mineral nutrients from sewage effluent. The pH of the water rose as high as 10.2, and it was estimated that between 60 and 100 g of carbon dioxide were absorbed from the atmosphere per m^2 per month at the height of the summer season. Invasion of carbon dioxide into the water from the atmosphere is thus sufficient to support high rates of photosynthesis, and Schindler (1971) concluded that carbon is unlikely to limit the standing crop of phytoplankton in almost any situation. Limitation, if it occurs, could not result in complete cessation of growth and is unlikely to be the cause of the ending of the spring outburst. It may be noted that Talling *et al.* (1973) found nearly the maximum algal crop to be expected on theoretical grounds in two soda lakes in Ethiopia, which had remarkably high reserves of carbon dioxide (51 to 67 m. equiv. $HCO_3^- + CO_3^{2-}$ per liter).

Organic Growth Factors

Many plankton algae are known to have requirements for substances such as thiamine, vitamin B_{12}, and biotin (Droop, 1962*b*; Provasoli, 1963, 1971), and it is possible that the concentrations of such substances in natural waters may sometimes be limiting. These and other organic growth factors are normally present in trace amounts in rain water (Parker and Wachtel, 1971), lake waters (Hutchinson, 1957), and sea waters (Provasoli, 1963; Duursma, 1965; Carlucci *et al.*, 1969). Thiamine, which appears to be produced mainly by bacteria, shows seasonal variation in concentration in freshwaters, decreasing during the increase of phytoplankton populations and increasing during their decay (Hagedorn, 1971). Vitamin B_{12} shows a seasonal variation in concentration in the sea, that in summer being extremely low; and it is lower in concentration in the open sea as compared with coastal waters (Provasoli, 1963). There has been disagreement as to whether this vitamin is ever a limiting factor. Droop at one time (in Oppenheimer, 1966) maintained that the lowest concentration found in seawater is still sufficient to provide for dense blooms but based this conclusion on determinations of the effect of vitamin B_{12} concentration on final yields. Effects on relative growth may be quite different. Provasoli, on the other hand, has urged the view that vitamin B_{12} may be an important limiting factor in production, and both Goldman (in Oppenheimer, 1966), with freshwater phytoplankton, and McLaughlin (in Oppenheimer, 1966), with marine phytoplankton, have found that small additions of vitamin B_{12} have stimulatory effects on photosynthesis as measured by the ^{14}C method. Carlucci and Silbernagel (1969) determined half-saturation constants (K_s) for the effects of vitamin B_{12} on *Cyclotella nana*, vitamin B_1 (thiamine) on *Monochrysis lutheri*, and biotin on *Amphidinium carteri*. From these results it appears unlikely that the vitamin concentrations in coastal waters would be limiting for phytoplankton but this would not necessarily be so in the open ocean. Nevertheless, as we have seen (p. 41) the relation of growth rate to vitamin B_{12} concentration is complex, and, furthermore, there is the uncertainty as to how much of the vitamin measured by bioassay in a natural water is available to phytoplankton species: much of it could be in bound form and unavailable to non-phagotrophic species (for references see Droop, 1962*b*; Provasoli, 1963). On the whole although vitamin concentrations may limit the growth of individual species (see p. 129), it does not seem likely that they could limit the overall growth of a mixed phytoplankton population in the spring outburst. It may be noted that Lund (1950; Lund *et al.*, 1963) did not take account of organic growth factors in his otherwise remarkably comprehensive studies with *Asterionella*. Hagedorn (1971) observed a decrease in thiamine concentration in the water during the development of an *Asterionella* bloom but did not present any evidence that this alga has a requirement for the vitamin.

Autoinhibitors

As far as I am aware, there is no strong evidence of production by a phytoplankton species of an autoinhibitor in sufficient concentration to bring growth to a standstill. Considering the relatively low population densities which are usually involved, the eventuality of this happening seems remote.

Loss in Outflow

The factors that have been considered so far are all such as may operate in batch cultures. In addition there is a factor, analogous to the wash-out which occurs in chemostat cultures, which is important under natural conditions. Thus, there is usually a flow of water through a lake and consequently a loss of phytoplankton in the outflow. A sudden flood may therefore wash a plankton bloom out, but where the retention time is long, the loss is negligible. For *Asterionella* in Windermere, Lund (1950) concluded that, during the spring maximum, loss in the outflow is more or less compensated for by the nutrients imported in the inflow.

Buoyancy

As we saw in the previous chapter, the sinking rate of phytoplankton cells may play a determining role in their biological success. A possible cause for the ending of the spring outburst might be that the progressive stabilization of the water-column and the decrease in viscosity of the water, as the season advances and the water becomes warmer, results in species adapted to grow under the more turbulent and colder conditions of early spring becoming unable to maintain themselves near the water surface. There seems to be no evidence that this is so for *Asterionella* (see Fig. 21) but an example of a plankton alga whose growth period is curtailed through sinking is provided by *Melosira italica* sub-species *subarctica*. Lund (1954, 1955) has shown that it sinks rapidly, at 3 to 5 times the rate of *Asterionella formosa*, so that decrease in numbers in the water occurs as turbulence diminishes. The diatom disappears almost completely from the lake while it is stratified (Fig. 34). *M. italica* is consequently most abundant in the lakes of the English Lake District between autumn and late spring. The filaments on settling out pass into the deposits and remain there until resuspended. Lund has shown experimentally that a proportion of the cells can remain alive, but not growing, under anaerobic conditions in the dark for as long as three years. Provided that sufficient nutrients are available, a summer maximum of this diatom may be induced by artificial destratification in a lake in which it is normally only abundant in winter (Lund, 1971).

Cells may not necessarily continue to sink until they reach the bottom. Steele and Yentsch (1960) showed that a maximum in phytoplankton chlorophyll at the bottom of the photic zone was due to a decreased sinking rate in this region. They argued that the cells themselves increase their buoyancy when they reach dark, nutrient-rich waters and carried out experiments showing that this actually does occur with *Skeletonema costatum* in laboratory culture.

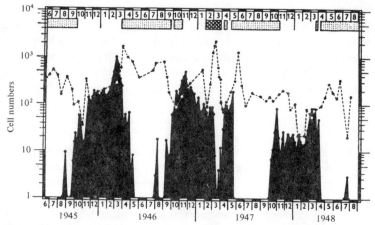

Figure 34. The abundance of *Melosira italica* sub-species *subarctica* on the deposits and in the plankton of Blelham Tarn, 1945–48. Live cells per ml in the 0–5 water column, solid black; in unit volume of deposit at 13 m interrupted line, both plotted on a logarithmic scale. Periods of stratification are indicated by the rectangles at the top: dotted, direct stratification; checkered, inverse stratification under ice. From J. W. G. Lund, The seasonal cycle of the plankton diatom, *Melosira italica* (Ehr.) Kutz. subsp. *subarctica* D. Müll., *J. Ecol.* (1954), *42*:156, fig. 2 (Blackwell Scientific Publications Ltd., Oxford, Eng.).

Grazing by Animals

This may have great effects on phytoplankton increase, as the following hypothetical illustration of Harvey (1945) shows. If 100 cells per liter undergo 6 successive cycles of division, the final population would be expected to be 6,400 cells per liter. If, however, 1 cell in every 10 is eaten in the intervals between divisions, the population will reach but 3,400 cells per liter, although only 413 cells have been eaten.

It is a necessary condition for the occurrence of the spring phytoplankton outbursts that, at the outset, the numbers of animals grazing on the algae should be negligible. There is a delay in zooplankton increase until the algae reach the threshold density necessary for the animals' reproduction, and the rate of multiplication is slow as compared with that of the algae. The effect of grazing is therefore likely to be appreciable only in the later stages of spring growth. In cold waters, in which the development of herbivores is extremely slow, the maximum of phytoplankton is evidently not controlled by grazing (Cushing,

1959*a*). In temperate waters the herbivores develop more quickly, and their numbers become sufficient toward the end of the spring growth period for them to have an appreciable effect on the course of phytoplankton increase. Cushing (1959*b*) and Steel (1963), on the basis of mathematical analysis of the course of phytoplankton increase in relation to probable controlling factors, consider that grazing is the most important factor bringing about the cessation of spring growth in waters around Britain, but as Riley (1963) points out, this may not be true in other regions. However, Smayda (1973) has shown that grazing by the copepod *Acartia clausi*, tintinnids, rotifers, and an amoeboid organism tentatively classified as an ebridian, was a major factor causing the decline of the spring bloom of *Skeletonema costatum* in Narragansett Bay, Rhode Island. McAllister (1970) pointed out that failure to take account of the difference in effects of nocturnally and continuously grazing copepods may introduce serious errors into calculation of production. He also put forward a theoretical model which predicts changes in plant and animal populations starting from different initial values of stock and the relationships of zooplankton ration and plant mortality to concentrations of organisms in the two trophic levels. In fresh water, grazing is, perhaps, of less importance. Lund (1959) is of the opinion that it has no appreciable effect on the spring increase of *Asterionella formosa*. In the English lakes the main development of zooplankton follows that of the phytoplankton. There is no evidence that the animals eat *Asterionella*, and it seems that their principal foods are bacteria and protozoa, which develop at the expense of the dying phytoplankton. Nauwerck (1963) has similarly concluded for Lake Erken in Sweden that phytoplankton is of secondary importance as a direct source of food for zooplankton. It is to be expected, of course, that nannoplankton are grazed to a greater extent than comparatively large forms such as *Asterionella*, but the difficulty of recognizing their remains makes direct evidence of this hard to obtain.

Parasitization

This has been suggested as a factor controlling the growth of marine phytoplankton, but there appears to be little definite evidence to support this view (ZoBell, 1946). One of the few fungal parasites of marine phytoplankton that has been reported is a not fully identified

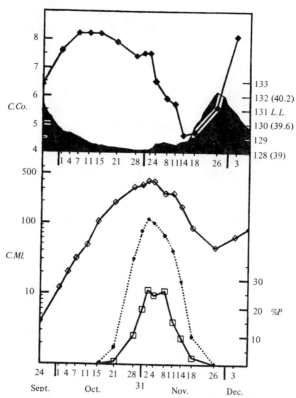

Figure 35. The relationship between numbers of a parasite, *Rhizophidium plank-tonicum*, and its host, *Asterionella formosa*, in Esthwaite Water, autumn 1946. *C.Co.*, average number of live cells per *Asterionella* colony (a solid line with black diamonds); *C.Ml.*, number of live *Asterionella* cells per ml (a solid line with white diamonds) and the number of live *Rhizophidium* cells per ml (a dotted line with black circles), both plotted on a logarithmic scale; *L.L.*, Windermere, North Basin lake level in feet (meters in parentheses) shown solid black; *%P*, percentage of *Asterionella* cells infected by living *Rhizophidium* cells (a solid line with white squares). From H. M. Canter and J. W. G. Lund, Studies on plankton parasites. I. Fluctuations in the numbers of *Asterionella formosa* Hass. in relation to fungal epidemics, *New Phytol.* (1948), 47:243, fig. 1 (Blackwell Scientific Publications Ltd., Oxford, Eng.).

form found as a heavy infestation on *Chaetoceros constrictus*, but not on other species of the genus, in a collection from Loch Carron, Scotland (Ingold, 1971). Many freshwater algae, however, are known to be parasitized by fungi. *Asterionella formosa* is attacked by the chytridaceous fungus *Rhizophidium planktonicum*. This parasite is nearly always present when the host is, but for most of the year its frequency is too low to reduce appreciably the numbers of *Asterionella* in the English lakes, and it seems that, although it may affect the course of the spring increase, it is rarely the cause of its end. The course of an epidemic is shown in Figure 35; such occurrences are most frequent in autumn and winter (Canter and Lund, 1948). So far it has not proved possible to grow *Rhizophidium* in artificial culture or to correlate the onset of epidemics with any single factor or group of factors. Viruses, cyanophages, and myxobacteria, causing lysis of blue-green algae, have been described and might be the cause of otherwise inexplicable disappearance of populations of these algae. The use of these pathogens in the control of blooms has been suggested (Fogg *et al.*, 1973).

There is often nothing corresponding to the stationary phase, as usually observed in laboratory cultures, following the cessation of the spring growth of phytoplankton. This is attributable to the effects of grazing and sedimentation in rapidly depleting a population which is not actively multiplying and to the killing, by high light intensities, of cells suffering from deficiency of certain nutrients. This evidently happens with *Asterionella formosa*, cells of which can continue to grow in the absence of sufficient supplies of silica, a process which results in self-destruction. The dead cells become covered with a mass of bacteria and sink rapidly, forming a flocculent deposit on the lake bottom (Lund *et al.*, 1963).

The most thorough studies of growth of phytoplankton in the spring are undoubtedly those of Lund, and as this chapter is concluded, it is worth repeating a point that he has constantly emphasized, namely, that his conclusions hold good only for *Asterionella formosa* in the English Lake District, and that for other species and for the same species in other parts of the world, the factors controlling growth may be quite different.

Some Other Aspects
of Phytoplankton Periodicity

During the summer period, following the spring maximum, there is normally a period of two months or so in temperate lakes and seas when the standing crop of phytoplankton remains at a relatively low and steady level. The low level of the standing crop is attributable to the depletion of nutrients from the photic zone by the sedimentation of the greater part of the cells produced in the spring maximum, thermal stratification stabilizing the water column so that there is no replenishment by mixing from below. The comparative stability in the amount of phytoplankton is attributable to the achievement of approximate equilibrium between it and the herbivore population, the algal reproductive rate being more or less balanced by the loss through grazing (Cushing, 1959a). A similar situation is characteristic of tropical lakes and seas throughout the year. These conditions are quite unlike those in a culture of limited volume and more resemble those in a chemostat culture (Riley, in Oppenheimer, 1966). Although nutrient concentrations in the water are low, there is continual regeneration by means of the zooplankton and bacteria, so that, as in the chemostat, growth is limited by the rate of supply of a nutrient in limiting amount. The rate of turnover of a nutrient element may be surprisingly high. Watt and Hayes (1963) have estimated the turnover times in inshore waters off

106

Halifax, Nova Scotia, as 1.5 days for dissolved inorganic phosphorus, 2.0 days for particulate phosphorus, and 0.5 days for dissolved organic phosphorus.

Such an equilibrium may be disturbed by changes in hydrographical conditions. Circulation and mixing may occur in tropical waters, but generally somewhat irregularly. Because of the greater change in density per degree at higher temperatures, a given amount of cooling produces much more active convection currents in a tropical water than in a temperate water. Continued cooling may thus lead to turnover, without the necessity of wind action, and a consequent enrichment with nutrients from below. Lake Victoria, which has moderately high densities of phytoplankton and a high rate of production throughout the year (Talling, 1961b), seems to owe its fertility to the efficient mixing brought about by seiche action in its extensive but shallow basin. A semi-tropical part of the Sargasso Sea, one of the few parts of the open sea in which the seasonal cycle and annual variability of primary productivity have been studied, has special hydrographical features (Ryther, 1963). The most important of these is a permanent thermocline at 400–500 m, above which a summer thermocline develops at about 100 m. Phytoplankton production remains at a low and fairly constant level throughout the year except when winter temperatures fall low enough for the upper thermocline to break down so that mixing of the water occurs down to the lower thermocline. The amount of nutrients thus made available is small but sufficient to produce an increase in phytoplankton, which gradually builds up to a maximum in the spring (see p. 129). In mild winters thermal stratification persists in the upper layers of water, and the production is correspondingly less.

Upwelling may be highly localized. Thus, in the Baie des Chaleurs, Canada, high phytoplankton productivity occurs in an area about 30 km in diameter in which cold nutrient-rich water is brought to the surface by a cyclonic gyre. East of the gyre, in the Gulf of St. Laurence, production is much less (Legendre and Watt, 1970). Even where upwelling occurs over an extensive area, its effects may be far from uniform. Strickland *et al.* (1969a) encountered two distinct types of situation in coastal waters off northern Chile and Peru. In one, "blue water," the water was rich in nutrients and had a small but actively growing standing stock of phytoplankton. The other, "brown water," had lower nutrient concentrations, a high standing stock of phytoplankton, and the higher productivity per unit area. The cause of these differences could not be established with certainty but it appeared probable that the phytoplankton in the blue water is kept in check by grazing, the effect of which was perhaps increased by dilution of the population with upwelling phytoplankton-free water.

High concentrations of plankton algae may not be the result of growth alone. From time to time there occur in the sea dense accumulations of phytoplankton known as "red tides," which not only are striking in themselves but also attract further attention by leading to mass mortality of marine invertebrates and fish. Most commonly, the organisms concerned are Dinophyceae, colored red by accumulation of carotenoids, but the term "red tide" has been extended to include phytoplankton blooms of other colors and composed of species belonging to other groups, *e.g.*, blue-green algae of the genus *Trichodesmium*. Since red tides appear to be characteristic of tropical waters and temperate waters in the summer, the phenomenon calls for some comment here. A typical example has been described by Paredes (1962) from the coast of Angola, where it is of fairly regular occurrence. From the air it was seen as vast parallel areas of reddish water. It persisted from 15 to 20 days and resulted in the death of large numbers of fish and crabs. The principal organism concerned was *Exuviella baltica* associated with other dinoflagellates and with diatoms. Red tides are usually confined to coastal waters, and undoubtedly one necessary condition is a high concentration of nutrients, as produced, for example, by upwelling or outflow of fresh water. There must also be paucity of zooplankton using the particular species for food and, possibly, particular conditions of light and temperature. The high population densities seem to be produced not directly by growth but by concentration at the surface as a result of wind on a lee shore, or current action, or downwelling (Ryther, 1955). The death of fish and other animals may sometimes be produced by mechanical clogging of their gills or by deoxygenation and production of substances such as hydrogen sulfide following decomposition of the phytoplankton, but certain red tide organisms, *e.g.*, *Gymnodinium veneficum*, are known to produce specific toxins (Abbott and Ballantine, 1957).

Freshwaters also sometimes produce red blooms of dinoflagellates, and in the Lago di Tovel in the Trentino region of Italy these are sufficiently striking and regular to provide a tourist attraction. However, the most familiar blooms in freshwaters are those of blue-green algae. The regulation of buoyancy by means of gas-vacuoles, which these algae possess, has already been discussed. Gas-vacuolate species forming large colonies rise or sink relatively rapidly, and it seems that sometimes they may overshoot the optimum depth for photosynthesis and, reaching inhibitory light intensities near the surface in which reduction in gas-vacuole content cannot be effected, become trapped. It is also possible that in periods of turbulence which do not permit stratification, the cells may become over-vacuolated so that they rise rapidly to the surface in a subsequent calm. Reynolds (1971) has shown that surface accumulations in the Shropshire meres, England, result from the concentration of colonies previously distributed in a water layer several meters deep, rather than from growth at the surface. These scums, which are often largely moribund, may be a considerable nuisance in freshwater and sometimes consist of toxin-producing strains, lethal to higher animals (Gorham, 1964; Fogg, 1969; Fogg et al., 1973).

The second maximum of phytoplankton which sometimes occurs in temperate waters in the autumn is perhaps principally due to augmentation of the nutrient supply brought about by increased circulation and mixing. Possibly a decline in the herbivore population may be a contributory factor (Cushing, 1959a). Although the nutrient supply is not limiting, phytoplankton growth is restricted by the diminishing light and the increased turbulence removing cells from the photic zone. These conditions may be considered as approximating those in the turbidostat type of continuous culture (Riley, in Oppenheimer, 1966). If the mixing rate is very great, the population in the photic zone will be depleted faster than it grows, and no autumn maximum will occur. This was evidently the situation in the autumn of 1947 in Windermere (Fig. 22).

The classic periodicity with spring and autumn maxima separated by a minimum in phytoplankton growth may be characteristic only of lakes in which the effects of nutrient depletion are dominant in summer. Round (1971) and Mathiesen (1971) have drawn attention to several instances of lakes in which, with increasing eutrophication, that is, over-enrichment with nutrients, the spring and autumn maxima tend to be obliterated by an extended mid-summer peak, the form of which seems to be determined by light and temperature.

While it is thus possible to suggest explanations in qualitative terms for the major features of phytoplankton periodicity, our knowledge will approach completeness only when a full quantitative account of the effects of various conditions can be given and used to predict the future course of plankton increase and decrease. It is not within the scope of this book to discuss in detail the mathematical models of plankton periodicity which have been put forward thus far, but perhaps an indication of the assumptions on which they are based and of their adequacy may be useful here. Riley (1963) has used the following general equation as the basis for his analysis:

$$\frac{dP}{dt} = P(P_h - R - G),$$

where P is the total phytoplankton population per unit of sea surface, P_h is a photosynthetic coefficient, R is a coefficient of phytoplankton respiration, and G is a grazing coefficient. P_h was estimated by an empirical equation from measurements of incident radiation, transparency of the water, depth of the mixed layer, and concentration of phosphate, which was taken as a measure of the general level of nutrients. Phytoplankton respiration was assessed from experimental data and assumed to increase exponentially with rise in temperature. The grazing coefficient was taken as proportional to the observed herbivore population. Several sets of data have been examined in this way by approximate integration over successive short periods of time during which the environmental conditions were assumed to remain constant. This gave curves showing the seasonal changes in relative terms. Three of these calculated curves, assigned arbitrary absolute values to give the best fit statistically with the observed phytoplankton populations, are shown together with the actual observed seasonal cycles in Figure 36. In spite of the many simplifying assumptions—the phytoplankton population, for example, is assumed to be homogeneous and the zooplankton to be non-selective in its grazing—the agreement between the calculated and observed curves is reasonably good. It must be remembered, however, that the situations chosen for this analysis were presumably selected for their comparative simplicity. Furthermore, besides the basic physical factors—incident radiation, temperature, and vertical eddy diffusivity—on which phytoplankton growth ultimately depends, others, such as transparency of the water, phosphate concentration, and herbivore numbers, which are dependent on

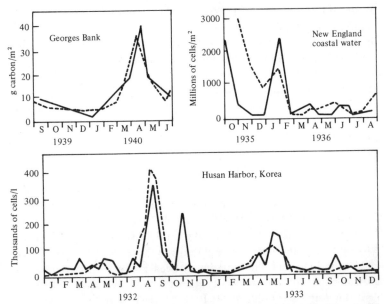

Figure 36. Comparison of observed seasonal cycles of phytoplankton (solid lines) with cycles calculated from a mathematical model. After G. A. Riley, Theory of food-chain relations in the ocean, in *The Sea,* ed. M. N. Hill, vol. 2 (1963), p. 442, fig. 1.

the phytoplankton population itself, were used. More fundamental approaches, such as those made by Riley and his collaborators (Riley, 1963), by Steele (1958), and by Cushing (1959*b*), inevitably give less satisfactory agreement between theoretical and observed values. However, the great value of these attempts is that they show where knowledge is inadequate and suggest methods and hypotheses for further investigation.

One line of further investigation is the experimental one. From what has already been said it will be appreciated that small-scale cultures, even if incubated *in situ* in the lake or the sea, are of limited use for this purpose and, if long continued, may give extremely misleading information. A deep tank (3 m diameter and 10 m deep), such as described by Strickland *et al.* (1969*b*), is useful for studying populations of single phytoplankton species under approximately natural conditions but, used over extended periods with natural populations, would become dominated by sessile forms growing on its walls. It is out of the question on grounds of expense to construct replicate artificial basins of sufficient size for the results of experiments conducted in them to have any great significance for the understanding of phytoplankton behavior. Water supply authorities rarely permit the use of their reservoirs for experimental purposes, and it is difficult to find natural bodies of water which can easily be divided into comparable portions, although this has been done in the case of Peter-Paul Lake in Wisconsin (Stross *et al.*, 1961). Strickland and Terhune (1961) were the first to experiment with plastic containers placed *in situ* for the study of phytoplankton behavior. Using a free-floating, 120 m³ capacity, transparent sphere, it was possible to study the growth and metabolism of a spring maximum type of marine phytoplankton bloom over a period of 24 days (Antia *et al.*, 1963). Valuable results were obtained but conditions were scarcely natural—the artificial stirring employed would have given different turbulence patterns to those arising from Langmuir circulation—and had the experiment been prolonged it is likely that growth of sessile algae on the container walls would have reduced the light intensity seriously. If such wall effects are to be negligible, enclosures have to be much larger. Lund (1972) has used vertical tubes of butyl rubber, 45.5 m in diameter, floating at the top and entering the mud at the bottom, in Blelham Tarn, English Lake District (Plate 4). In such tubes the chlorophyll in sessile algae growing on the walls is less than one thousandth of that in the plankton, and phytoplankton was found to remain qualitatively similar to that of the main lake for extended periods. However, the quantitative characteristics altered, and, although the general pattern of the annual periodicity remained the same as in the main lake, the spring and autumn maxima were lower, evidently because access to nutrients such as nitrate, phosphate, and silica, entering the lake from outside, was prevented. Lund (1972) points out that this suggests that once inflowing nutrients are removed, the recovery of a small lake from eutrophication may be rapid. Such tubes have great promise for the study of the effects of particular nutrients.

Another approach to more detailed understanding is through statistical studies of the relationship of the seasonal variations in phytoplankton productivity and biomass to simultaneously observed values of all the major factors which may be envisaged as affecting them. In such a

Plate 4. Blelham Tarn, English Lake District, with the inflated tops of two experimental tubes visible. By courtesy of J. W. G. Lund and A. E. Ramsbottom from J. W. G. Lund, Preliminary observations on the use of large experimental tubes in lakes, *Verh. int. Ver. Limnol.* (1972), *18*:71–77, fig. 1 (Schweizerbart'sche Verlagsbuchhandlung, Stuttgart).

study, on Lake Maggiore in Italy, Goldman *et al.* (1968) carried out stepwise multiple regression analysis with data obtained by measuring 28 variables at 9 depths on 14 days between May and December. This period covered the end of the spring maximum, the summer minimum, and a late summer maximum. A close relationship was found between primary productivity as measured by the radiocarbon technique and the solar radiation incident on the water surface during the period of the determination but the correlation was not perfect. Further analysis failed to relate the discrepancy to any one of the 27 other variables, including concentrations of nutrients such as nitrate, phosphate, and silicate. The effects of zooplankton grazing were not evaluated. It was clear that there was no simple connection between phytoplankton photosynthesis and phytoplankton biomass and that undetermined variables, such as the physiological state of the alga (one may guess that internal concentrations of elements such as nitrogen and phosphorus would be important), must have had considerable effects.

In most quantitative studies, however, it is usually assumed that the properties and responses of phytoplankton are unvarying. As we saw in Chapter IV, this is far from being so in cultures, and, although conditions in natural waters are not usually so extreme as they may be in the laboratory, we should expect something of the great variability in intensity and pattern of metabolism of which algae are capable to be exhibited in natural populations. Visual evidence that this happens is provided by those plankton species which accumulate large quantities of carotenoids (p. 58). *Botryococcus braunii*, for example, may be encountered in both the green and orange forms in lakes and presumably has the corresponding high and low rates of photosynthesis and growth observed in cultures.

Quantitative evidence that variation in rate of metabolism occurs in natural populations is provided by the estimates of the ratio of photosynthesis at light saturation to chlorophyll (assimilation number) which are available in the literature and which have been summarized by Platt and Subba Rao (1973). We have already seen (p. 58) that this ratio varies within wide limits in a not altogether intelligible manner in laboratory-grown algae. The situation is similar with naturally occurring phytoplankton, and there are the added complications that in estimations in natural waters it is difficult to make a distinction between chlorophyll in living algae and that in detritus and that mixed populations may include forms in which much of the light energy utilized in photosynthesis may be absorbed by pigments other than chlorophyll. Some workers, such as Steele and Baird (1961), have obtained fairly consistent values for assimilation numbers, but others, such as El-Sayed (1967), have recorded variation of over a hundredfold. These variations cannot at present be related to particular factors but on the face of it they suggest variations in metabolic activity of natural phytoplankton at least as great as that described for laboratory cultures in Chapter IV. Rodhe *et al.* (1958) found in Lake Erken that the ratio of carbon fixed in the 1–2 m layer to the corresponding amounts of chlorophyll remained at a rather constant low value during the spring growth but rose in the summer months to higher values, which showed no simple pattern of variation. Similarly Steele and Baird (1961) found that in two different areas of the North Sea, in the waters of which there did not appear to be any appreciable amount of "dead" chlorophyll, there were distinct seasonal trends in the assimilation number from about 1 in spring to about 2 in summer, with a decline again in autumn (Fig. 37). Taguchi (1970) found a similar trend in Akkeshi Bay, Hokkaido, the assimilation number being 0.83 in February, when nutrients were at a high level, but 6.7 in May, when the spring maximum was beginning to decline. These findings are in accord with the results of some laboratory experiments in which assimilation number was found to rise with nutrient deficiency but, as we saw (p. 58), other experi-

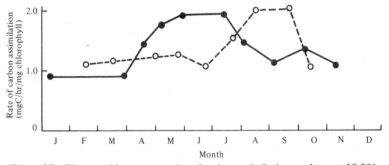

Figure 37.   The monthly average ratios of carbon assimilation per hour at 10,000 lux to chlorophyll concentration for two areas of the North Sea, Fladen, 0-30 m, solid line; Aberdeen Bay, dashed line. From J. H. Steele and I. E. Baird, Relations between primary production, chlorophyll and particulate carbon, *Limnol. oceanogr.* (1961), *6*:77, fig. 8.

ments have given a contrary indication, and Strickland (1965) concluded that nutrient limitation reduces the assimilation number in marine phytoplankton. These contradictions cannot at present be resolved. Evidence of trends in metabolism is also provided by the carbon/chlorophyll ratio. Steele (1962) observed a rise of about tenfold in this ratio in the open North Sea from spring to summer. This corresponded well with the observed depletion of nutrients and was of the same order of magnitude as that found by Steele and Baird (1962) in cultures.

Determinations of the elementary composition or the contents of major fractions such as carbohydrate, fat, and protein in phytoplankton populations might also be expected to give some indication of their metabolic characteristics. Few such analyses have been made on natural populations but those of Haug *et al.* (1973) are valuable because they were made over a complete growing season. Lipid content was found always to be below 10 per cent of the dry organic matter. The protein/carbohydrate ratio varied between 0.48 and 1.83. Low values could be attributed to a predominance of dinoflagellates with cellulose walls—just as in laboratory grown material (p. 60). The amounts of acid-soluble glucan in diatoms was found to vary between 7.7 and 36.5 per cent of organic dry matter, and changes in this fraction were the main cause of variation in the protein/carbohydrate ratio in samples consisting largely of diatoms. When variations arising from changes in floristic composition were allowed for, the trends were similar to those reported for batch cultures (Chapter IV). The spring bloom was characterized by a rapid decrease in protein/carbohydrate ratio, and thereafter this ratio fluctuated in a parallel manner to the nitrate concentration in the seawater, as one would expect. The nitrogen/phosphorus ratio remained remarkably constant throughout. It has generally been observed that this ratio tends to constancy and to be the same as that in the seawater in which the phytoplankton grew (Harvey, 1945). Again, this is to be expected; it follows from what we know of luxury consumption of nutrients (p. 42) that the amounts of these in the cells will reflect the concentrations available in the medium rather than the kinds of phytoplankton species present.

Finally, in this chapter, the possibility of short-term variations in growth and metabolism should be considered. It might be expected that the regular alternation of light and dark in the natural habitat would induce a considerable degree of synchronization of cell division and consequently a diurnal periodicity in metabolic activity. It is well established that pronounced diurnal rhythms of photosynthetic activity are shown by phytoplankton. Doty and Oguri (1957) measured radio-carbon uptake by samples of natural marine plankton from surface waters in the tropics for short periods under constant conditions. They found that the same population photosynthesized nearly 6 times as fast under such conditions at 8 A.M. as it did at 7 P.M. Similar behavior was found in samples from deep water. This diurnal fluctuation was found to decrease going north from the equator and ceased altogether at a latitude of 60° to 70° N. In accordance with these observations, Shimada (1958) found three- to fourfold diurnal variation at 18° N, and Yentsch and Ryther (1957) a twofold variation at Wood's Hole, 42° N. However, from data tabulated by Platt and Subba Rao (1973) it appears that this trend may not always be a regular one. In fresh waters in Germany, Ohle (1958, 1961) has observed decreases of about 20 per cent in photosynthetic activity between morning and afternoon.

The explanation of these variations is not obvious. Their magnitude indicates changes in rate of photosynthesis rather than of respiration, and they are accompanied by corresponding changes in chlorophyll (Yentsch and Ryther, 1957), although variation is still evident when rates per unit amount of chlorophyll are considered (Sournia, 1967). Steemann Nielsen and Jørgensen (1962) considered the effect to be due to a reduced rate of chlorophyll synthesis at high light intensities combined with grazing by zooplankton; actual destruction of chloro-phyll, it was pointed out, occurs only after photosynthesis has ceased altogether. Ohle (1961), who found no evidence of endogenous rhythms in the phytoplankton populations which he studied, showed that the variation is reduced if the turbulence in the water is increased. He postulated that the decreased rates of photosynthesis were brought about by accumulation of waste products within the cells. On the other hand, the occurrence of endogenous rhythms in phytoplankton under natural conditions is well established (Sweeny and Hastings, 1962; Eppley et al., 1971), so that one might reasonably expect to find diurnal rhythms contributing to the observed variations in photo-synthesis. If synchronization of cell division were to occur in lakes or the sea, we might expect to find Tamiya's D-type cells predominating in the morning and L cells predominating later on in the day. This would agree with the observed variations in photosynthesis and chlorophyll. The approximately 12 hours light/12 hours dark cycle prevailing in the tropics would be expected to produce a much greater degree of synchrony than the almost continuous daylight of high latitudes in the summer, and hence the metabolic variation in the former situation would be the most pronounced. These various explanations are not necessarily mutually exclusive, and it might well be that the observed fluctuations are due to a combination of effects.

V I I I

Phytoplankton Distribution and Seasonal Succession

It is a useful simplification in many kinds of investigation to regard phytoplankton as a single homogeneous entity, and it must be largely true that the total biomass of phytoplankton is determined by the physical and chemical factors in the environment and is not dependent on the species which are represented in it. Nevertheless, phytoplankton normally consists of a heterogeneous collection of organisms, and the problems posed by the distribution and seasonal succession of the species present are not only of interest in themselves, but, since qualitative differences may have effects on the higher components of the food chain, are of economic importance. The spatial distribution of phytoplankton species may be limited by barriers such as land or water masses of unfavorable temperature or salinity, but otherwise the operative factors are largely the same as those determining distribution in time or, in other words, succession. It should always be remembered that the changes observed in a fixed position are, to a greater or lesser extent, a combination of true succession and movement of water masses. For present purposes it is more appropriate to concentrate on the problems of seasonal succession.

The general character of phytoplankton succession during the course of the year may be illustrated by results obtained by Nauwerck (Rodhe

119

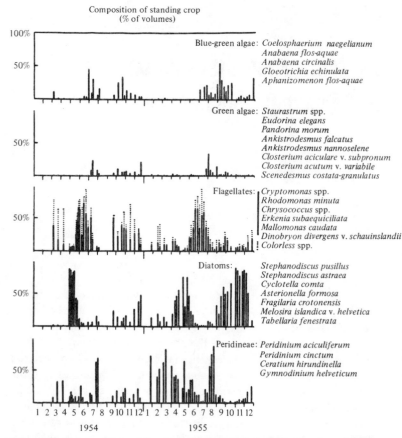

Figure 38. Composition of the phytoplankton in Lake Erken, southern Sweden, during the years 1954 and 1955. Amounts of the main groups are expressed in terms of percentages of the total fresh weight, and the principal species are listed. From W. Rodhe, R. A. Vollenweider, and A. Nauwerck, The primary production and standing crop of phytoplankton, in *Perspectives in Marine Biology,* ed. A. A. Buzzati-Traverso (1958), p. 304, fig. 4.

et al., 1958) for Erken, a eutrophic lake in central Sweden (Fig. 38). In both the years covered by these observations diatoms made up the larger part of the standing crop in April and May, being followed by flagellates, such as *Rhodomonas minuta* and *Dinobryon divergens,* and these by green and blue-green algae. Although there was a general similarity between the sequences in the two years, there was not exact repetition; the autumn diatom maximum in 1955, for example, was much more pronounced than that in 1954. Other examples of phytoplankton succession in lakes are given by Hutchinson (1967). Similar patterns occur in temperate seas, with diatoms usually predominating in spring and autumn and flagellates during the summer (Raymont, 1963). In tropical waters succession is less regular. Nannoplankton have usually been ignored in studies of succession but reports on their periodicity have been made by Lund (1961) for English lakes and Kristiansen (1971) for Danish lakes.

It will be obvious that the factors concerned in such a succession must be various and that the interactions between them are likely to be complex. The following seem to be important:

Temperature and Light

Different species undoubtedly have different temperature and light requirements, and it could be that a particular species predominates at a particular season because the prevailing temperature and light conditions favor it. It should be noted that light and temperature conditions occur in different combinations according to season: spring, high light and low temperature; summer, high light and high temperature; autumn, low light and high temperature; winter, low light and low temperature (see Strickland, 1965; Hutchinson, 1967). A particular species might be able to withstand a high light intensity at a low temperature but not at a high temperature and so would tend to occur in spring rather than summer. It is customary to designate plankton organisms as oligothermal, polythermal, or eurythermal, according to whether they appear to be cold-requiring, warmth-requiring, or able to tolerate a wide temperature range (Ruttner, 1953; Hutchinson, 1967). Some algae, such as the sea-ice diatom, *Fragilaria sublinearis,* studied by Bunt *et al.* (1966), seem to be genetically adapted to growth at low temperature and incapable of survival above a temperature as low as

10° C. However, the temperature at which an organism is most abundant in nature does not necessarily correspond to its optimum temperature as determined in the laboratory; the planktonic strain of *Chlorella pyrenoidosa* studied by Fogg and Belcher (1961) was isolated from Torneträsk in Swedish Lappland, in which the temperature does not rise above 7° C, yet its optimum in culture was found to be about 20° C (Fig. 8). The discrepancies and contradictions regarding the temperature requirements of *Skeletonema costatum*, a eurythermal species, are also disturbing (Braarud, 1962). The optimum temperature for growth in the sea has been variously reported as 12° to 13°C and 14° to 20°C, and Smayda (1973) found perceptible growth of this species in Narragansett Bay, Rhode Island, at temperatures ranging from 0° to 22°C. From data tabulated by Smayda (1973) it is evident that the optimum temperature for growth in culture, which has been reported by different workers as 20° and 30° C, depends very much on the culture medium as well as on other factors such as light intensity and adaptation. Strains of this, and other, species with different temperature requirements may exist and add further complication to the interpretation of observations. The occurrence of blue-green algae in certain situations may be directly attributable to their tolerance of high temperatures, but otherwise it seems that the abundance of a particular species can rarely be put down purely and simply to the temperature's being optimal for it. The absence of a species may, of course, be ascribable to the direct effect of unduly high or low temperatures. Similarly with the intensity and quality of light there are so many possibilities of adaptation and interaction with other factors that it seems unlikely that any of these *per se* can be the major factor determining the occurrence of a species at a particular time or place. Photoperiod, as we have seen (pp. 48, 117), has important effects on the life-cycles and metabolic activities of algae both in culture and in nature, and it seems likely that change in day-length with season could play a part in determining succession although this possibility does not appear to have been investigated.

Hydrographical Conditions

Hydrographical conditions undoubtedly may play an important part in determining the abundance of particular species at particular seasons. The effects of turbulence and the differences in buoyancy among species have already been discussed in Chapter VI, and it seems probable that the change from the turbulent conditions of early spring to the more stabilized water column in summer could favor the replacement of rapidly sinking forms by more buoyant or motile algae. This was demonstrated clearly by Moss (1969a) in a sheltered pond. The best example from a larger water body of the occurrence of a particular plankton species at particular times being determined by turbulence is that of *Melosira italica,* which has already been discussed (p. 101).

Concentration of Nutrients

The differing requirements of different species for nutrients may operate to give succession. An obvious example here is that when a diatom maximum has come to an end because of exhaustion of silica, enough other nutrients may yet be available to support further growth of algae, such as Chlorophyceae, which do not require silica. Other features of succession might be the result of changes in concentration in other nutrients and shifts in ion ratios occurring in the water as the season advances. It seems a sad but inescapable conclusion that much of the painstaking investigation of the effects of these factors on growth of phytoplankton in culture is of little use in interpreting the events in lakes or the sea. Some workers have used final yield as their measure of the effect of nutrient concentration on growth, whereas effects on lag phase and the relative growth constant are undoubtedly of more importance in determining succession. Apart from this, it has not been sufficiently realized how much the level of other factors—light intensity, temperature, presence of organic chelating agents, etc.—may determine the effect of a given concentration of a nutrient or that the relative growth rate of an alga is more directly determined by the intracellular concentration of a particular nutrient than by its concentration in the surrounding medium.

A promising approach is through the kinetics of nutrient uptake as determined in laboratory experiments. The two constants in the expression relating uptake to concentration (p. 41), u_∞, the maximum rate, and K_c, the half-saturation constant, might be used to predict the growth rate of a given species under particular sets of conditions and so explain its occurrence at a particular time and place (Strickland, 1972). Thus, Eppley *et al.* (1969) found that small oceanic species had the lowest K_c values for nitrate (about 0.5 μg-at. N per liter) and coastal dinoflagellates the highest (5 to 10 μg-at. N per liter), and this fits in with the general observation that oceanic waters are poorer in nitrate than coastal ones. Eppley (1970) attempted to use this kinetic approach to predict species distributions but could not fully account in these terms for the success of rock-pool species or of red tide organisms in the vicinity of La Jolla, California. Smayda (1973) also used this approach to explain the periodicity of *Skeletonema costatum* in Narragansett Bay, Rhode Island, but concluded that it was premature to attempt rigorous application of uptake and growth kinetics based on laboratory studies. It must be remembered that several factors may be limiting at the same time so that the polynomial expression devised by Droop (1973; see p. 42) would be more appropriate than the simple Monod equation used by Eppley. Furthermore, uptake is not a simple function of concentration but, as we have seen (p. 71), is dependent on sinking rate and extent of vertical migration, both of which vary with species and with a given species according to its physiological condition.

Observation of succession in relation to the changes in the chemistry of natural waters which actually take place offers another approach. A notable investigation along these lines was that of Pearsall (1932), who carried out a study of the composition of the phytoplankton in relation to dissolved substances in the English lakes. Some of his conclusions were as follows: (1) diatoms increase when the water is richest in phosphate, nitrate, and silica; (2) *Dinobryon divergens*, in hard-water lakes, develops maxima when the silica content falls below 5 ppm and, in all lakes, is favored by a rise in the ratio of nitrate to phosphate; (3) desmids and other green algae occur mainly during the summer depletion of nutrients, the former being favored by a low calcium content and a low nitrate/phosphate ratio; (4) the abundance of blue-green algae is correlated with high concentrations of dissolved organic matter in the water, these algae being able to increase at very low concentrations of inorganic nutrients, such as nitrate and phosphate.

As we saw in Chapter VI, Pearsall's conclusions regarding diatom growth have been substantiated by subsequent work. His inferences regarding *Dinobryon* (2, above), however, were found by Hutchinson (1944, 1967) not to hold for lakes in other regions. The appearance of this alga in Linsley Pond, Connecticut, is correlated with a rise in the nitrate/phosphate ratio but is independent of variations in dissolved silica, although in other lakes it may be correlated with falling silica concentration but not with the nitrate/phosphate ratio. Rodhe (1948) considered that *Dinobyron* is favored by reduction in phosphate concentration, but it has been grown in culture in phosphate-rich media (Talling, 1962). Pearsall's conclusion (3, above) that low calcium concentrations favor green algae, and in particular desmids, also needs modification. Chu (1942) in culture experiments found that whereas *Pediastrum boryanum* grows best with low concentrations of calcium, the concentration of this element in nearly all fresh waters is within the optimum range for *Staurastrum paradoxum.* Moss (1973c), using relative growth rate as his criterion, concluded that there is no evidence for the view that high concentrations of calcium depress the growth of desmids characteristic of oligotrophic waters. It appears, rather, that the available free carbon dioxide is the determining factor. Species characteristic of oligotrophic waters seem to be confined to utilizing the undissociated forms and therefore do not grow at pH values above 8.6 to 8.85, whereas species characteristic of eutrophic waters may grow at above pH 9.0, either using bicarbonate directly or using undissociated carbon dioxide at very low concentrations (Moss, 1973a). The importance of carbon dioxide and pH in determining the occurrence of algal species in particular waters is a matter of controversy (Goldman, 1973; Shapiro, 1973), but it seems improbable that seasonal changes in either of these factors could be sufficiently great in a given body of water to affect the succession of phytoplankton. Chu (1943) also found that no increase or decrease of the nitrate/phosphate ratio markedly affected the growth of any of the plankton algae which he studied, so long as the concentration of these radicals remained within the optimum range for the organisms in question.

The correlation of abundance of blue-green algae with the content of dissolved organic matter found by Pearsall (4, above) seems to hold generally, but the basis for this is still obscure (Fogg, 1969; Fogg *et al.,* 1973). Most planktonic blue-green algae which have been grown in culture have been found to do well in mineral media devoid of organic supplements other than a chelating agent to maintain iron in an available form (Gerloff *et al.,* 1950; Staub, 1961). Some blue-green algae are capable of growing in the dark on organic substrates, but most are obligate phototrophs (Fogg *et al.,* 1973). Rodhe (1948) found that *Gloeotrichia echinulata* requires an unidentified thermolabile organic growth factor, but so far there is no other evidence that planktonic blue-green algae, whether freshwater or marine, have requirements for vitamins. Additions of thiamine to natural populations *in situ* in a lake in large plastic cylinders either had no effect upon the numbers of Cyanophyceae or actually caused them to decrease (Hagedorn, 1971). Only certain benthic marine species have been shown to require vitamin B_{12} (van Baalen, 1962). Although Pearsall's correlation of abundance of blue-green algae with the dissolved organic content of the water a month earlier suggests the latter as the causative factor, it is a well-established characteristic of these algae in culture to liberate substantial quantities of extracellular organic substances (Fogg, 1952), but it does not seem likely that the relatively high concentrations of dissolved organic matter usually found in the waters in which they occur can be produced mainly by the blue-green algae themselves. The explanation of the correlation is probably not simple but depends on a number of factors. By forming complexes many types of organic substance may regulate the inorganic environment, perhaps reducing the toxicity of some ions or making others more available (Fogg and Westlake, 1955). Blue-green algae generally photosynthesize and fix nitrogen more rapidly at low oxygen concentrations (Stewart and Pearson, 1970), and these are more likely to be attained in waters containing high concentrations of organic matter. It is also possible that by photoassimilation of organic substances planktonic blue-green algae may be able to achieve greater growth at low light intensities than they would if carbon dioxide were the only carbon source (Fogg, 1969; Fogg *et al.*, 1973).

Pearsall's second point about blue-green algae—that they are able to develop at very low concentrations of inorganic nutrients—has been confirmed by Hutchinson (1967). This does not necessarily contradict Vollenweider's (1968) generalization that planktonic blue-green algae tend to be abundant in water bodies in which nutrient concentrations exceed 10 mg P per m^3 and 200 to 300 mg N per m^3 in the spring and/or the specific supply loading per unit area reaches 0.2 to 0.5 g P per m^2 per year and 5 to 10 g N per m^2 per year. Blue-green algae are present in these waters at seasons when nutrient concentrations are lowest. In part, this may be explained by the ability of some species to fix free nitrogen, N_2. Planktonic blue-green algae possessing heterocysts, such as species of the genera *Anabaena*, *Aphanizomenon*, and *Gloeotrichia*, are able to fix nitrogen, and certain non-heterocystous forms, such as *Plectonema* and *Oscillatoria* spp., may also do so under microaerophilic conditions, but the common plankton species *Microcystis aeruginosa* apparently does not (Stewart, 1973; Fogg *et al.*, 1973). *In situ* determinations of nitrogen fixation in lakes have shown that the rate is almost invariably correlated with the abundance of heterocystous species and is light dependent (Horne and Goldman, 1972; Fogg *et al.*, 1973). Figure 39 shows the development of a maximum of *Anabaena* with a parallel increase in nitrogen fixation at a time when concentrations of ammonia and nitrate in the water were low. A rise in concentration of these two forms of combined nitrogen later suppressed nitrogen fixation but not the growth of the blue-green algae. Several non-planktonic marine species of blue-green algae have been isolated in pure culture and have been shown to be nitrogen-fixing. There is no clear evidence that marine planktonic forms can fix nitrogen, although there are strong indications that *Trichodesmium*, which sometimes forms dense blooms in tropical waters, possesses the property (Dugdale *et al.*, 1961; Stewart, 1971). In summary it may be said that an ability to fix nitrogen may sometimes be the reason for the predominance of blue-green algae at times when shortage of nitrogen limits other plankton algae, but it cannot be a general explanation of their position in the seasonal succession, nor, of course, does it provide any explanation for the observation that they are most abundant when the concentration of phosphate is particularly low.

The paradox that planktonic blue-green algae develop most massively when inorganic nutrients are in shortest supply cannot be completely resolved at present but two points may be advanced in tentative explanation. First, blue-green algae have remarkable powers of accumulating intracellular reserves of phosphate, in the form of volutin, and of nitrogen, in the form of the pigment phycocyanin and the unusual polypeptide, containing only arginine and aspartic acid, cyanophycin (Fogg *et al.*, 1973). Such reserves might suffice for several cycles of cell division after the blue-green algae had exhausted the external supply. Secondly, abundance of blue-green algae has usually been compared with average or surface concentration of nutrients whereas, as we have seen, populations of these algae are often highly localized at particular depths. The role of chemical stratification in determining the vertical distribution of photosynthetic bacteria is well recognized (Moss, 1969a) and something similar may occur with blue-green algae. Additionally, it

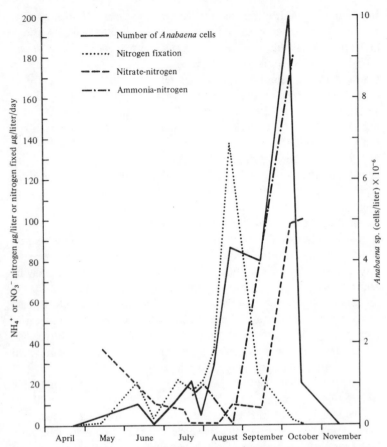

Figure 39. Changes in concentrations of ammonia- and nitrate-nitrogen, nitrogen fixation, and numbers of *Anabaena* in the surface water of Sanctuary Lake, Pennsylvania, during 1959. Replotted data from V. A. Dugdale and R. C. Dugdale, Nitrogen metabolism in lakes. II. Role of nitrogen fixation in Sanctuary Lake, Pennsylvania, *Limnol. Oceanogr.* (1962), 7:170.

may well be that at a time when the concentration of nutrients in surface water is low, the blue-green algae because of their vertical migration may enjoy access to deeper waters with ample nutrient supply (Fogg, 1969; Fogg *et al.*, 1973).

Less information is available about the rôle of nutrient concentrations in the succession in marine plankton. However, an apparently clear-cut instance of the dependence of a diatom maximum on the concentration of vitamin B_{12} in the waters of the Sargasso Sea has been reported by Menzel and Spaeth (1962). Using the diatom *Cyclotella nana* to assay for vitamin B_{12}, the authors found the concentration in the surface 50 m to vary from 0.03 mμg/liter down to an undetectable amount from May to October. *Coccolithus huxleyi*, which does not require vitamin B_{12}, is dominant in the phytoplankton under these conditions. The bloom of diatoms, with *Rhizosolenia stolterfothii* and *Bacteriastrum delicatulum* predominating, occurred in April when the level of B_{12} was increased to 0.06 to 0.1 mμg/liter, evidently by mixing in of deeper water. This brings us to the possibility, envisaged by Hutchinson (1944) and others, that traces of organic growth-promoting or growth-inhibiting substances produced by the algae themselves may play an important part in succession.

Growth-Promoting and Growth-Inhibiting Substances

Some evidence for the presence of growth-promoting and growth-inhibiting substances in natural waters and their rôle in determining the total biomass of phytoplankton has already been discussed (p. 99). Clearly they may be produced by specific organisms and may have specific effects on particular species, so that their rôle in interspecific competition and determining the qualitative composition of the phytoplankton may be considerable. These biologically active substances may be liberated during healthy growth or after the death of the cells, perhaps only during decomposition under particular conditions.

There is mounting evidence for the production of growth-promoting substances by algae in culture. Lewin (1958) detected thiamine in filtrates from cultures of *Coccomyxa* sp. Bednar and Holm-Hansen (1964) found *Coccomyxa* to release biotin, and Nakamura and Gowans (1964) showed that a *Chlamydomonas* sp. liberated extracellular nicotinic acid. Among important marine phytoplankton species *Skeletonema costatum* and *Stephanopyxis turris* (vitamin B_{12} requirers) released both thiamine and biotin in culture; *Gonyaulax polydra* (vitamin B_{12} requirer) released thiamine, and *Coccolithus huxleyi* (thiamine requirer) released both vitamin B_{12} and biotin (Carlucci and Bowes, 1970a). The rate of vitamin liberation was found to vary with the species of alga, the concentration of vitamin which it itself required, and the age of the culture. During exponential growth, vitamin was liberated as an extracellular product of healthy cells but in the stationary phase some was liberated by cell lysis. Carlucci and Bowes concluded that a major proportion of the vitamins dissolved in natural seawater is contributed by the phytoplankton. Plant hormones of the auxin type have been demonstrated in extracts of *Chlorella* and various plankton algae, in filtrates from cultures of *Anabaena cylindrica,* and from lake water containing a nearly unialgal growth of *Oscillatoria* sp. (Bentley, 1958, 1960).

Many workers have reported that certain species inhibit the growth of others in mixed culture (see Hutchinson, 1967; Sieburth, 1968; Fogg, 1971; Tassigny and Lefèvre, 1971, for summaries). Vance (1965), for example, showed by quantitative culture techniques that the blue-green algae, *Microcystis aeruginosa,* produces a factor inhibitory to the growth of other algae and that it itself is vulnerable to a substance liberated by *Euglena* sp. Perhaps some of these effects are due to specific antibiotic substances, but since the conditions of culture have not always been strictly controlled nor the substance isolated and chemically characterized, there is considerable uncertainty. Proctor (1957), however, showed clearly that inhibition of *Haematococcus pluvialis* by *Chlamydomonas reinhardtii* is due to a fatty acid liberated on the death of the *Chlamydomonas* cells. Jakob (1961) found that *Nostoc muscorum* produces in bacteria-free culture a "dihydroxy-anthraquinone," which inhibits growth of other algae such as *Cosmarium, Phormidium,* and *Euglena.* On the other hand, in a careful study of the growth of two planktonic diatoms, *Asterionella formosa* and *Fragilaria crotonensis,* in mixed culture and in media containing culture filtrates, Talling (1957) found no evidence for the production by either species of any extracellular substance which appreciably modified the growth of the other. There is a great deal of evidence that some algae produce antibacterial substances (see Jørgensen, 1962; Duff *et al.,* 1966). While these may have no direct effect on other algae, they may obviously indirectly affect competition among algal species by their effects on associated bacteria.

Complex interactions between species were found by Carlucci and Bowes (1970*b*) in experiments with mixed cultures, following up their work, described above, on vitamin production. It was demonstrated that a vitamin liberated by one species would support the growth of another species requiring that vitamin; thus, *Dunaliella tertiolecta* and *Skeletonema costatum* produced utilizable thiamine for *Coccolithus huxleyi*. However, release of toxic materials or vitamin inactivators also occurred (Fig. 40). The intensity of the growth-promoting and growth-inhibiting effects depended on the species present and the conditions of growth. In systems with more than two algal components the beneficial effects on the vitamin utilizers often lasted for a brief period only.

It seems likely that some of these effects observed in laboratory culture occur in the natural environment. Zavarzina (1959) has reported the presence of a factor inhibitory to the growth of *Scenedesmus quadricauda* in various reservoir and lake waters. This factor, which could be adsorbed on activated charcoal, appeared to originate from other algae rather than from the lake mud. Smayda (1963), however, has put forward reasons for thinking that antibiosis plays no very important part in phytoplankton succession, and in view of the ready adaptation of some algae to antibiotics (see p. 31) this seems probable. Extracellular products of certain algae may neutralize the effects of toxic substances. Whitton (1967) showed that the non-dialyzable extracellular products of *Anabaena cylindrica* reduce the toxicity of polymixin both towards *A. cylindrica* itself and to the other blue-green algae, *Chlorogloea fritschii* and *Anacystis nidulans*. Johnston (1963b) likewise concluded from his experimental studies that the quality of sea waters was primarily a matter of their content of growth-promoting substances rather than of inhibitors. These growth-promoting substances appeared to have different effects on different species, since bioassays of various seawater samples, enriched with phosphate and nitrate and containing adequate amounts of vitamin B_{12}, gave different results according to whether the assay organism was *Skeletonema costatum* or *Peridinium trochoideum*.

Hagedorn (1971) carried out experiments in which thiamine was added to give a concentration of 10 μg per liter to the water in plastic containers, 1 m diameter and 10 m deep, immersed in the surface of a lake. Within three days the population of Chlorococcales had increased to nearly 900 per cent as compared with the control, whereas the corresponding percentages were 100 for *Trachelomonas,* 70 for pennate diatoms, and 55 for *Closterium*. Repeat experiments gave similar results. Pratt (1966) described the competition between the diatom *Skeletonema costatum* and the xanthophycean *Olisthodiscus luteus* in

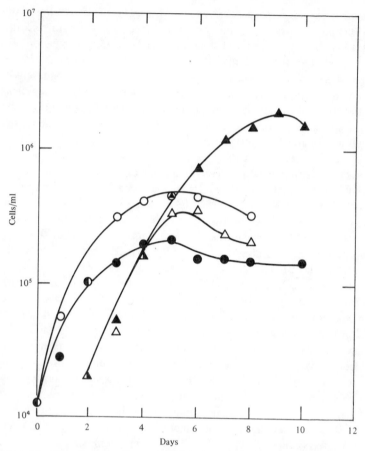

Figure 40.   Growth of vitamin-starved cells of the flagellate *Coccolithus huxleyi* added to cultures of the diatom *Skeletonema costatum* after a preincubation of 2 days. ○, *S. costatum* control; △, *C. huxleyi* control; ●, *S. costatum* with *C. huxleyi* (inhibitory effect); ▲, *C. huxleyi* with *S. costatum* (growth-promoting effect). From A. F. Carlucci and P. M. Bowes, Vitamin production and utilization by phytoplankton in mixed culture, *J. Phycol.* (1970), 6:393–400, fig. 2 B (Phycological Society of America Inc.).

Narragansett Bay, Rhode Island, and was able to relate his observations to effects demonstrated in cultures. Media prepared with filtrates from *S. costatum* cultures, fortified with nutrients, supported normal growth of *O. luteus* with no enhancing or inhibiting effects. Dominance of *S. costatum* over *O. luteus in situ* appeared to be due to faster growth of the former. On the other hand, the growth of *S. costatum* was affected in media prepared with *O. luteus* culture filtrate as a base; small aliquots of this filtrate enhanced growth while larger proportions were inhibitory. This perhaps explains why *S. costatum* maxima occur before, between, and after, but never during, the *O. luteus* maxima in Narragansett Bay, which may rise to 9500 cells per ml. The inhibitory substance excreted by *O. luteus* has a molecular weight of less than 2000 (Stuart, quoted by Smayda, 1973).

These results suggest that traces of biologically active substances play an important part in determining succession but it appears that the interactions are so complex that full understanding is still a long way off.

Selective Grazing

If herbivores graze selectively, they may have important effects on succession. Obviously there must be some discrimination according to size, but cells may also be selected or rejected on a more subtle basis. Among the best evidence for selective grazing is that obtained by Edmondson (1964) from a study of the relation of egg production by rotifers in the English lakes to numbers of various phytoplankton algae and bacteria. Data from a total of 400 samples obtained over a period of two years were subjected to statistical analysis. The partial correlation coefficients obtained between egg production and numbers of a phytoplankton species presumably are a measure of the extent to which the alga was eaten, although, of course, there may be effects due to differential digestion and the chemical composition of the food organisms. The results given in Table 7 for *Keratella cochlearis* indicate that this rotifer seeks out and eats *Chrysochromulina*. *Rhodomonas,* an alga about the same size as *Chrysochromulina,* was eaten to a much smaller extent, although it was nearly twice as abundant. Other organisms, *e.g., Stichococcus* and miscellaneous flagellates, were also selected, but *Chlorella* was perhaps avoided.

TABLE 7

Relation of reproductive rate of *Keratella cochlearis* to temperature and abundance of specific food organisms in four lakes. Temperature in ° C, abundance of food organisms in μg dry wt/liter (Edmondson, 1964)

Independent variable	Mean value of independent variable	Partial correlation coefficient
Temperature	11.11	0.416
Chlorella	6.33	−0.007
Stichococcus	1.63	0.129
Coccomyxa	5.09	0.023
Chlamydomonas	2.32	0.013
Chrysochromulina	26.52	0.415
Miscellaneous flagellates	13.84	0.152
Colorless flagellates	18.97	0.052
Rhodomonas	49.48	0.135
Chrysococcus	7.52	0.009

Similar results were obtained for *Kellicottia longispina,* whereas a third rotifer, *Polyarthra vulgaris,* showed a high correlation with *Cryptomonas* but little correlation with anything else except temperature. These findings are in general agreement with the conclusions which Nauwerck (1963) drew from his study of the relations between zooplankton and phytoplankton in Lake Erken. Nauwerck found that, although Peridineae, diatoms, and blue-green algae were most abundant in the phytoplankton, small chrysomonads were most important, both qualitatively and quantitatively, as food for the animals. Cryptomonads came next and small green algae and small diatoms were considerably less important.

It has been generally supposed that feeding of marine zooplankton is unselective (see Raymont, 1963) but observations by Conover (1966) on *Calanus hyperboreus* show that after having been fed on small particles this copepod is slow in adapting its pattern of feeding for the capture of large particles. This suggests that this species might reduce the numbers of nannoplankton relative to larger forms, if nothing more. Another copepod, *Acartia clausi,* has been shown to graze selectively on *Skeletonema costatum* (Smayda, 1973).

Selective grazing evidently occurs under natural conditions and must exert some influence on phytoplankton succession, but it remains to be demonstrated that this influence is ever the predominating one.

There is thus no evidence that any one factor is of over-riding importance in determining the abundance of particular species at particular seasons, and it seems probable that in each instance a complex of interacting factors is concerned. An interesting approach to dealing with this situation has been made by Margalef (1958), based mainly on his studies of phytoplankton in the Gulf of Vigo. Three stages in succession may be distinguished:

1. A phase of growth, characterized by algae, *e.g.*, small diatoms, which have small cells (and consequently a high surface/volume ratio), a preference for high concentrations of nutrients, and a relatively high rate of growth (one or two divisions per day). These algae are easily grown in crude culture.

2. A mixed community of forms, which often have bigger cells and lower relative growth rate and which are not so easily grown in culture, *e.g.*, larger diatoms.

3. A mixture of forms, many of them motile, characterized by being extremely difficult to culture and therefore presumably having complex nutritional requirements, *e.g.*, dinoflagellates, particularly red-tide organisms. Irregularities in horizontal distribution become most pronounced at this stage, with patches "changing from moment to moment like clouds in the sky."

This is perhaps a picture of general validity for temperate waters. For example, in the English lakes stage 1 would seem to be represented by *Asterionella, Melosira,* and *Chlorella,* stage 2 by *Tabellaria, Staurastrum,* and *Mallomonas,* and stage 3 by dinoflagellates and blue-green algae. Johnston (1963*a*) has commented that the phytoplankton succession in waters off the northeast coast of Scotland resembles in principle that described by Margalef and has made the interesting observation that species characteristic of stage 1, *e.g., Skeletonema* and *Thalassiosira,* are sensitive to antimetabolites such as sulfanilamide and benzimidazole, whereas later forms, *e.g., Chaetoceros* spp. and *Rhizosolenia alata,* are most resistant.

Margalef determined the index of diversity,

$$d = \frac{S-1}{\log_e N},$$

where S is the number of species and N the number of individuals, at different intervals during the growing season. The index d was found to vary in a rather regular manner during the succession (Fig. 41), being between 1 and 2 at the beginning and increasing up to 5 in stage 3. Ignatiades (1969) observed a similar seasonal drift of species diversity in Saronicos Bay, Aegean Sea, Goldman *et al.* (1968) found the same trend in the freshwater Lake Maggiore, and Moss (1973*d*) also found it in three freshwater bodies. Since the determination of diversity indices is extremely laborious, Margalef (1968) has sought for simpler methods. Pigment diversity seems to be an adequate expression of the complexity of photosynthetic systems, and he has shown a statistically significant positive correlation between the ratio of absorbances of pigment extracts at 430 and 665 nm (an index of the ratio of carotenoids to chlorophylls) and the species diversity index in certain situations. The

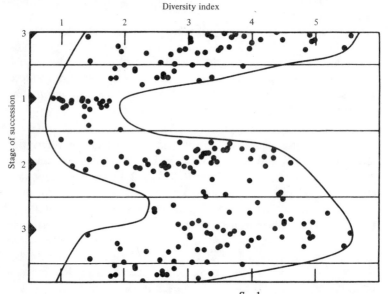

Figure 41. Variation of diversity index (d = $\dfrac{S-1}{\log_e N}$) in relation to stage of succession in a number of spot samples taken in surface waters of the Gulf of Vigo, summer, 1955. From R. Margalef, Temporal succession and spatial heterogeneity in phytoplankton, in *Perspectives in Marine Biology*, ed. A. A. Buzzati-Traverso (1958), p. 338, fig. 5.

seasonal drift of this pigment index in natural waters resembles that which occurs in mixed laboratory cultures. However, as Moss (1973d) has pointed out, the correlation may be fortuitous since the absorbance ratio reflects amounts, rather than kinds, of carotenoids and can be affected by degradation of chlorophylls to phaeo-pigments.

The general picture seems to be that at the beginning of the growing season the water is rich in nutrients and behaves rather like an ordinary artificial culture medium containing only inorganic constituents—a point we have noted before (p. 84). The most rapidly growing species then becomes dominant. As a result of this growth, mineral nutrients are depleted, and the less labile organic by-products of metabolism begin to accumulate. Consequently, the species which now succeed are not the most rapidly growing but those which are best able to make use of, or tolerate, the organic matter. In such a community we might expect interrelations among species to be complex and the number of species consequently high. Increased stability of the water column allows chemical stratification so that the number of available niches is increased. Reduction of mixing at this stage also results in reduced rates of growth and permits discrete clouds of organisms to be formed and to persist. Flagellates can remain in the surface waters, whereas diatoms sink so that the heterogeneity is vertical as well as horizontal. Persistence of remnants of populations which reached their peaks earlier in the year may also contribute to species diversity in the summer (Moss, 1973c). Some analogy with succession in terrestrial communities may be seen, as here, too, pioneer vegetation tends to consist of a few rapidly growing species, which are followed by a more complex and integrated community with more species. However, more definite biochemical information about the changes in dissolved organic matter and studies in culture of the nutritional requirements of the organisms concerned are needed to fill in the details of this picture.

References / Index

References

Aach, H. G. 1952. Über Wachstum und Zusammensetzung von *Chlorella pyrenoidosa* bei unterschiedlichen Lichtstärken und Nitratmengen. *Arch. Mikrobiol. 17:*213–246.

Abbott, B. C., and Ballantine, D. 1957. The toxin from *Gymnodinium veneficum* Ballantine. *J. mar. biol. Ass. U.K. 36:*169–189.

Al-Hasan, R. H., Coughlan, S. J., Pant, A., and Fogg, G. E. In the press. Seasonal variations in phytoplankton and glycollate concentrations in the Menai Straits, Anglesey. *J. mar. biol. Ass. U.K.*

Allen, M. B., French, C. S., and Brown, J. S. 1960. Native and extractable forms of chlorophyll in various algal groups. In *Comparative Biochemistry of Photoreactive Systems,* ed. M. B. Allen; New York and London; p. 33–52.

Antia, N. J., and Kalmakoff, J. 1965. Growth rates and cell yields from axenic mass culture of fourteen species of marine phytoplankton. In *Fisheries Research Board of Canada. Manuscript Report Series (Oceanographical and Limnological).* No. 203. Vancouver, B.C.

Antia, N. J., McAllister, C. D., Parsons, T. R., Stephens, K., and Strickland, J. D. H. 1963. Further measurements of primary production using a large-volume plastic sphere. *Limnol. Oceanogr. 8:*166–183.

Bainbridge, R. 1957. The size, shape and density of marine phytoplankton concentrations. *Biol. Rev. 32:*91–115.

Bassham, J. A., and Calvin, M. 1957. *The Path of Carbon in Photosynthesis.* Englewood Cliffs, N.J.

Bednar, T. W., and Holm-Hansen, O. 1964. Biotin liberation by the lichen alga *Coccomyxa* sp. and by *Chlorella pyrenoidosa. Pl. Cell Physiol., Tokyo 5:*297–303.

141

Beklemishev, C. W., Petrikova, M. N., and Semina, G. I. 1961. On the cause of the buoyancy of plankton diatoms. *Trudȳ Inst. Okeanol. 51:*33–36 (in Russian).

Belcher, J. H., and Miller, J. D. A. 1960. Studies on the growth of Xanthophyceae in pure culture. IV. Nutritional types amongst the Xanthophyceae. *Arch. Mikrobiol. 36:*219–228.

Belser, W. L. 1963. Bioassay of trace substances. In *The Sea*, vol. 2, ed. M. N. Hill; New York and London; p. 220–231.

Bentley, J. A. 1958. Role of plant hormones in algal metabolism and ecology. *Nature, Lond. 181:*1499–1502.

Bentley, J. A. 1960. Plant hormones in marine phytoplankton, zooplankton and seawater. *J. mar. biol. Ass. U.K. 39:*433–444.

Berman, T., and Rodhe, W. 1971. Distribution and migration of *Peridinium* in Lake Kinneret. *Mitt. int. Verein. theor. angew. Limnol. 19:*266–276.

Bernard, F. 1963. Density of flagellates and Myxophyceae in the heterotrophic layers related to environment. In *Symposium on Marine Microbiology*, ed. C. H. Oppenheimer; Springfield, Ill.; p. 215–228.

Billaud, V. A. 1967. Aspects of the nitrogen nutrition of some naturally occurring populations of blue-green algae. In *Environmental Requirements of Blue-green Algae*. Corvallis, Oreg.; p. 35–53.

Bongers, L. H. J. 1956. Aspects of nitrogen assimilation by cultures of green algae. *Meded. LandbHoogesch. Wageningen. 56:*1–52.

Braarud, T. 1962. Species distribution in marine phytoplankton. *J. oceanogr. Soc. Japan*, 20th Anniversary vol., p. 628–649.

Brook, A. J., Baker, A. L., and Klemer, A. R. 1971. The use of turbidometry in studies of the population dynamics of phytoplankton populations with special reference to *Oscillatoria agardhii* var. *isothrix. Mitt. int. Verein. theor. angew. Limnol. 19:*244–252.

Bunt, J. S. 1969. Observations on photoheterotrophy in a marine diatom. *J. Phycol. 5:*37–42.

Bunt, J. S., Owens, O., Van, O. H., and Hoch, G. 1966. Exploratory studies on the physiology and ecology of a psychrophilic marine diatom. *J. Phycol. 2:*96–100.

Canter, H. M., and Lund, J. W. G. 1948. Studies on plankton parasites. I. Fluctuations in the numbers of *Asterionella formosa* Hass. in relation to fungal epidemics. *New Phytol. 47:*238–261.

Caperon, J., and Meyers, J. 1972a. Nitrogen-limited growth of marine phytoplankton. I. Changes in population characteristics with steady-state growth rate. *Deep Sea Res. 19:*601–618.

Caperon, J., and Meyers, J. 1972b. Nitrogen-limited growth of marine phytoplankton II. Uptake kinetics and their role in nutrient limited growth of phytoplankton. *Deep Sea Res. 19:*619–632.

Carlucci, A. F., and Bowes, P. M. 1970a. Production of vitamin B_{12}, thiamine and biotin by phytoplankton. *J. Phycol. 6:*351–357.

Carlucci, A. F., and Bowes, P. M. 1970b. Vitamin production and utilization by phytoplankton in mixed culture. *J. Phycol. 6:*393–400.

Carlucci, A. F., and Silbernagel, S. B. 1969. Effect of vitamin concentrations on growth and development of vitamin requiring algae. *J. Phycol. 5:*64–67.

Carlucci, A. F., Silbernagel, S. B., and McNally, P. M. 1969. Influence of temperature and solar radiation on persistence of vitamin B_{12}, thamine, and biotin in seawater. *J. Phycol. 5:*302–305.

Cassie, R. M. 1963. Microdistribution of plankton. *Oceanogr. & mar. Biol. a. Rev. 1:*223–252.

Chu, S. P. 1942. The influence of the mineral composition of the medium on the growth of planktonic algae. I. Methods and culture media. *J. Ecol. 30:*284–325.

Chu, S. P. 1943. The influence of the mineral composition of the medium on the growth of planktonic algae. II. The influence of the concentration of inorganic nitrogen and phosphate phosphorus. *J. Ecol. 31:*109–148.

Collyer, D. M., and Fogg, G. E. 1955. Studies on fat accumulation by algae. *J. exp. Bot. 6:*256–275.

Conover, R. J. 1966. Feeding on large particles by *Calanus hyperboreus* (Kroÿer). In *Some Contemporary Studies in Marine Science,* ed. H. Barnes; London; p. 187–194.

Conway, K., and Trainor, F. R. 1972. *Scenedesmus* morphology and flotation. *J. Phycol. 8:*138–143.

Costlow, J. D., Jr. 1971. *Fertility of the Sea,* 2 vols., ed. J. D. Costlow, Jr.; New York; 622 pp.

Cushing, D. H. 1959a. The seasonal variation in oceanic production as a problem in population dynamics. *J. Cons. perm. int. Explor. Mer 24:*455–464.

Cushing, D. H. 1959b. On the nature of production in the sea. *Fishery Invest., Lond., Ser. 2, 22*(6).

Danforth, W. F. 1962. Substrate assimilation and heterotrophy. In *Physiology and Biochemistry of Algae,* ed. R. A. Lewin; New York and London; p. 99–123.

Davis, C. O., Harrison, P. J., and Dugdale, R. D. 1973. Continuous culture of marine diatoms under silicate limitation. I. Synchronized life cycle of *Skeletonema costatum. J. Phycol. 9:*175–180.

Von Denffer, D. 1948. Über einen Wachstumshemmstoff in alternden Diatomeenkulturen. *Biol. Zbl. 67:*7–13.

Dinsdale, M. T., and Walsby, A. E. 1972. The interrelations of cell turgor pressure, gas-vacuolation and buoyancy in a blue-green alga. *J. exp. Bot. 23:*561–570.

Doty, M. S., and Oguri, M. 1957. Evidence for a photosynthetic daily periodicity. *Limnol. Oceanogr. 2:*37–40.

Dring, M. J. 1971. Light quality and the photomorphogenesis of algae in marine environments. In *Fourth European Marine Biology Symposium,* ed. D. J. Crisp; New York; p. 375–392.

Droop, M. R. 1954. Conditions governing haematochrome formation and loss in the alga *Haematococcus pluvialis* Flotow. *Arch. Mikrobiol. 20:*391–397.

Droop, M. R. 1961a. Vitamin B_{12} and marine ecology: the response of *Monochrysis lutheri. J. mar. biol. Ass. U.K. 41:*69–76.

Droop, M. R. 1961b. Some chemical considerations in the design of synthetic culture media for marine algae. *Botanica Mar. 2:*231–246.

Droop, M. R. 1962a. On cultivating *Skeletonema costatum:* some problems. In *Beiträge zur Physiologie und Morphologie der Algen.* Vorträge aus dem Gesamtgebiet der Botanik, Deutschen Botanischen Gesellschaft, N.F. *1:*77–82.

Droop, M. R. 1962b. Organic micronutrients. In *Physiology and Biochemistry of Algae,* ed. R. A. Lewin; New York and London; p. 141–159.

Droop, M. R. 1966. Vitamin B_{12} and marine ecology. III. An experiment with a chemostat. *J. mar. biol. Ass. U.K. 46:*659–671.

Droop, M. R. 1968. Vitamin B_{12} and marine ecology. IV. The kinetics of uptake, growth and inhibition in *Monochrysis lutheri. J. mar. biol. Ass. U.K. 48:*689–733.

Droop, M. R. 1969. Algae. In *Methods in Microbiology,* vol. 3B., ed. J. R. Norris and D. W. Ribbons; London and New York; p. 269–313.

Droop, M. R. 1973. Nutrient limitation in osmotrophic protista. *Am. Zoologist 13:*209–214.

Duff, D. C. B., Bruce, D. L., and Antia, N. J. 1966. The antibacterial activity of marine planktonic algae. *Can. J. Microbiol. 12:*877–884.

Dugdale, R. C., Menzel, D. W., and Ryther, J. H. 1961. Nitrogen fixation in the Sargasso Sea. *Deep Sea Res. 7:*297–300.

Dugdale, V. A., and Dugdale, R. C. 1962. Nitrogen metabolism in lakes. II. Role of nitrogen fixation in Sanctuary Lake, Pennsylvania. *Limnol. Oceanogr. 7:*170–177.

Duursma, E. K. 1965. The dissolved organic constituents of sea water.

In *Chemical Oceanography*, vol. 1, ed. J. P. Riley and G. Skirrow; London and New York; p. 433–475.

Ebata, T., and Fujita, Y. 1971. Changes in photosynthetic activity of the diatom *Phaeodactylum tricornutum* in a culture of limited volume. *Pl. Cell Physiol., Tokyo 12:*533–541.

Eberly, W. R. 1967. Problems in the laboratory culture of planktonic blue-green algae. In *Environmental Requirements of Blue-green Algae.* Corvallis, Oreg.; p. 7–34.

Edmondson, W. T. 1964. Reproductive rate of planktonic rotifers as related to food and temperature in nature. *Ecol. Monogr. 35:*61–111.

El-Sayed, S. Z. 1967. On the productivity of the southwest Atlantic Ocean and the waters west of the Antarctic Peninsula. In *Antarctic Research Series 11 (Biology of the Antarctic Seas III);* ed. G. A. Llano and W. L. Schmitt; Washington, D.C.; p. 15–47.

Eppley, R. W. 1970. Relationships of phytoplankton species distribution to the depth distribution of nitrate. *Bull. Scripps Instn Oceanogr. non-tech. Ser. 17:*43–49.

Eppley, R. W., Holmes, R. W., and Strickland, J. D. H. 1967. Sinking rates of marine phytoplankton measured with a fluorometer. *J. exp. mar. Biol. Ecol. 1:*191–208.

Eppley, R. W., Holm-Hansen, O., and Strickland, J. D. H. 1968. Some observations on the vertical migration of dinoflagellates. *J. Phycol. 4:*333–340.

Eppley, R. W., Packard, T. T., and MacIsaac, J. J. 1970. Nitrate reductase in the Peru current. *Mar. Biol. 6:*195–199.

Eppley, R. W., Rogers, J. N., and McCarthy, J. J. 1969. Half-saturation constants for uptake of nitrate and ammonium by marine phytoplankton. *Limnol. Oceanogr. 14:*912–920.

Eppley, R. W., Rogers, J. N., McCarthy, J. J., and Sournia, A. 1971. Light/dark periodicity in nitrogen assimilation of the marine phytoplankters *Skeletonema costatum* and *Coccolithus huxleyi* in N-limited chemostat culture. *J. Phycol. 7:*150–154.

Eppley, R. W., and Strickland, J. D. H. 1968. Kinetics of marine phytoplankton growth. In *Advances in Microbiology of the Sea,* ed. M. R. Droop and E. J. Ferguson Wood; London and New York; p. 23–62.

Eppley, R. W., and Thomas, W. H. 1969. Comparison of half-saturation constants for growth and nitrate uptake of marine phytoplankton. *J. Phycol. 5:*375–379.

Erwin, J., and Bloch, K. 1964. Biosynthesis of unsaturated fatty acids in micro-organisms. *Science, Washington, D.C. 143:*1006–1012.

Fay, P., and Kulasooriya, S. A. 1973. A simple apparatus for the continuous culture of photosynthetic micro-organisms. *Br. phycol. J. 8:*51–57.

Findlay, I. W. O. 1972. Effects of external factors and cell size on the cell division rate of a marine diatom *Coscinodiscus pavillardii* Forti. *Int. Revue ges. Hydrobiol. 57:*523–533.

Fitzgerald, G. P. 1968. Detection of limiting or surplus nitrogen in algae and aquatic weeds. *J. Phycol. 4:*121–126.

Fitzgerald, G. P., and Nelson, T. C. 1966. Extractive and enzymatic analyses for limiting or surplus phosphorus in algae. *J. Phycol. 2:*32–37.

Fogg, G. E. 1944. Growth and heterocyst production in *Anabaena cylindrica* Lemm. *New Phytol. 43:*164–175.

Fogg, G. E. 1949. Growth and heterocyst production in *Anabaena cylindrica* Lemm. II. In relation to carbon and nitrogen metabolism. *Ann. Bot., Lond., N.S. 13:*241–259.

Fogg, G. E. 1952. The production of extracellular nitrogenous substances by a blue-green alga. *Proc. R. Soc., B 139:*372–397.

Fogg, G. E. 1956. Photosynthesis and formation of fats in a diatom. *Ann. Bot., Lond., N.S. 20:*265–285.

Fogg, G. E. 1959. Nitrogen nutrition and metabolic patterns in algae. *Symp. Soc. exp. Biol. 13:*106–125.

Fogg, G. E. 1969*a*. The physiology of an algal nuisance. *Proc. R. Soc., B 173:*175–189.

Fogg, G. E. 1969*b*. Survival of algae under adverse conditions. *Symp. Soc. exp. Biol. 23:*123–142.

Fogg, G. E. 1971. Extracellular products of algae in freshwater. *Arch. Hydrobiol. Beih. Ergebn. Limnol. 5:*1–25.

Fogg, G. E. 1973. Phosphorus in primary aquatic plants. *J. int. Ass. Wat. Poll. Res. 7:*77–91.

Fogg, G. E. In the press *a*. Biochemical pathways in unicellular plants. In *The Functioning of Photosynthetic Systems in Different Environments;* ed. J. P. Cooper; New York.

Fogg, G. E. In the press *b*. Primary Productivity. In *Chemical Oceanography,* vol. 2, ed. J. P. Riley and G. Skirrow; London and New York.

Fogg, G. E., and Belcher, J. H. 1961. Physiological studies on a planktonic μ-alga. *Verh. int. Verein. theor. angew. Limnol. 14:*893–896.

Fogg, G. E., Eagle, D. J., and Kinson, M. E. 1969. The occurrence of glycollic acid in natural waters. *Verh. int. Verein. theor. angew. Limnol. 17:*480–484.

Fogg, G. E., Smith, W. E. E., and Miller, J. D. A. 1959. An apparatus for the culture of algae under controlled conditions. *J. biochem. microbiol. Technol. Engng 1:*59–76.

Fogg, G. E., Stewart, W. D. P., Fay, P., and Walsby, A. E. 1973. *The Blue-green Algae.* London and New York; 459 pp.

Fogg, G. E., and Than-Tun. 1960. Interrelations of photosynthesis and assimilation of elementary nitrogen in a blue-green alga. *Proc. R. Soc., B 153:*111–127.

Fogg, G. E., and Westlake, D. F. 1955. The importance of extracellular products of algae in freshwater. *Verh. int. Verein. theor. angew. Limnol. 12:*219–232.

Fuhs, G. W. 1969. Phosphorus content and rate of growth in the diatoms *Cyclotella nana* and *Thalassiosira fluviatilis. J. Phycol. 5:*312–321.

Fujimoto, Y., Iwamoto, H., Kato, A. and Yamada, K. 1956. Studies on the growth of *Chlorella* by continuous cultivation. *Bull. agric. chem. Soc. Japan 20:*13–18.

Ganf, G. G. 1969. Physiological and ecological aspects of the phytoplankton of Lake George Uganda. Ph.D. diss., University of Lancaster, England.

Gerloff, G. C., and Fishbeck, K. A. 1969. Quantitative cation requirements of several green and blue-green algae. *J. Phycol. 5:*109–114.

Gerloff, G. C., Fitzgerald, G. P., and Skoog, F. 1950. The isolation, purification, and nutrient solution requirements of blue-green algae. In *Symposium on the culturing of algae;* ed. J. Brunel, G. W. Prescott, and L. H. Tiffany; Dayton, Ohio; p. 27–44.

Gerloff, G. C., and Skoog, F. 1954. Cell contents of nitrogen and phosphorus as a measure of their availability for growth of *Microcystis aeruginosa. Ecology 35:*348–353.

Gerloff, G. C., and Skoog, F. 1957. Nitrogen as a limiting factor for the growth of *Microcystis aeruginosa* in southern Wisconsin lakes. *Ecology 38:*556–561.

Gimmler, H., Ullrich, W., Domanski-Kaden, J., and Urbach, W. 1969. Excretion of glycolate during synchronous culture of *Ankistrodesmus braunii* in the presence of disalicylidene-propanediamine or hydroxypyridinemethanesulfonate. *Pl. Cell Physiol., Tokyo 10:*103–112.

Goldberg, E. D. 1963. The oceans as a chemical system. In *The Sea,* vol. 2, ed. M. N. Hill; New York and London; p. 3–25.

Goldman, C. R. 1960. Molybdenum as a factor limiting primary productivity in Castle Lake, California. *Science, Washington, D.C. 132:*1016–1017.

Goldman, C. R. 1961. Primary productivity and limiting factors in Brooks Lake, Alaska. *Verh. int. Verein. theor. angew. Limnol. 14:*120–124.

Goldman, C. R. 1965. *Primary Productivity in Aquatic Environments,* ed. C. R. Goldman. *Memorie Ist. ital. Idrobiol.* 18 Suppl.

Goldman, C. R., Gerletti, M., Javornicky, P., Melchiorri-Santaclini, U., and Amezaga, E. de. 1968. Primary productivity, bacteria, phyto and zooplankton in Lake Maggiore: correlations and relationships with ecological factors. *Memorie Ist. ital. Idrobiol. 23:*48–127.

Goldman, J. C. 1973. Carbon dioxide and pH: effect on species succession of algae. *Science, Washington, D.C. 182:*306–307.

Gorham, E. 1957. The chemical composition of some waters from lowland lakes in Shropshire, England. *Tellus 9:*174–179.

Gorham, P. R. 1964. Toxic algae. In *Algae and Man,* ed. D. F. Jackson; New York; p. 307–336.

Gran, H. H., and Braarud, T. 1935. A quantitative study of the phytoplankton in the Bay of Fundy and the Gulf of Maine (including observations on hydrography, chemistry and turbidity). *J. biol. Bd Can. 1:*279–467.

Gross, F., and Zeuthen, E. 1948. The buoyancy of plankton diatoms: a problem of cell physiology. *Proc. R. Soc., B 135:*382–389.

Haeckel, E. 1890. *Plankton-Studien.* Jena.

Hagedorn, H. 1971. Experimentelle Untersuchungen über den Einfluss des Thiamins auf du natürliche Algenpopulation des Pelagials. *Arch. Hydrobiol. 68:*382–399.

Harder, R. 1917. Ernährungsphysiologische Untersuchungen an Cyanophyceen, hauptsächlich dem endophytischen *Nostoc punctiforme. Z. Bot. 9:*145–242.

Hardy, A. C. 1956. *The Open Sea: the World of Plankton.* London; 335 pp.

Hart, T. J. 1966. Some observations on the relative abundance of marine phytoplankton populations in nature. In *Some Contemporary Studies in Marine Science,* ed. H. Barnes; London; p. 375–393.

Harvey, H. W. 1945. *Recent advances in the chemistry and biology of sea water.* Cambridge.

Hase, E. 1962. Cell division. In *Physiology and Biochemistry of Algae,* ed. R. A. Lewin; New York and London; p. 617–624.

Haug, A., Myklestad, S., and Sakhaug, E. 1973. Studies on the phytoplankton ecology of the Trondheimsfjord. I. The chemical composition of phytoplankton populations. *J. exp. mar. Biol. Ecol. 11:*15–26.

Hellebust, J. A. 1970. The uptake and utilization of organic substances by marine phytoplankton. In *Organic Matter in Natural Waters,* ed. D. W. Hood; Fairbanks; University of Alaska Institute of Marine Science Occasional Publications, no. 1; p. 225–256.

Hellebust, J. A. 1971. Glucose uptake by *Cyclotella cryptica:* dark induction and light inactivation of transport system. *J. Phycol.* 7:345–349.

Hill, M. N., ed. 1963. *The Sea,* vol. 2; New York; 554 pp.

Hinshelwood, C. N. 1946. *The chemical kinetics of the bacterial cell.* Oxford.

Hobbie, J. E., and Wright, R. T. 1965. Competition between planktonic bacteria and algae for organic solutes. *Memorie Ist. ital. Idrobiol.* 18 Suppl., p. 175–185.

Holmes, R. W., and Anderson, G. C. 1963. Size fractionation of [14] C-labeled natural phytoplankton. In *Symposium on Marine Microbiology,* ed. C. H. Oppenheimer; Springfield, Ill.; p. 241–250.

Holm-Hansen, O. 1971. Determination of microbial biomass in deep ocean profiles. In *Fertility of the Sea,* vol. 1, ed. J. D. Costlow, Jr.; New York; p. 197–207.

Holm-Hansen, O., Lorenzen, C. J., Holmes, R. W., and Strickland, J. D. H. 1965. Fluorometric determination of chlorophyll. *J. Cons. perm. int. Explor. Mer 30:*3–15.

Holm-Hansen, O., Nishida, K., Moses, V., and Calvin, M. 1959. Effects of mineral salts on short-term incorporation of carbon dioxide in *Chlorella. J. exp. Bot. 10:*109–124.

Holsinger, E. C. T. 1955. The distribution and periodicity of the phytoplankton of three Ceylon lakes. *Hydrobiologia 7:*25–35.

Hood, D. W., ed. 1970. *Organic Matter in Natural Waters.* Fairbanks; University of Alaska Institute of Marine Science Occasional Publications, no. 1; 625 pp.

Hoogenhout, H., and Amesz, J. 1965. Growth rates of photosynthetic microorganisms in laboratory cultures. *Arch. Mikrobiol. 50:*10–24.

Horne, A. J., Fogg, G. E., and Eagle, D. J. 1969. Studies *in situ* of the primary production of an area of inshore Antarctic sea. *J. mar. biol. Ass. U.K. 49:*393–405.

Horne, A. J., and Goldman, C. R. 1972. Nitrogen fixation in Clear Lake, California. I. Seasonal variation and the role of heterocysts. *Limnol. Oceanogr. 17:*678–692.

Horne, A. J., and Wrigley, R. C. Hunting phytoplankton by remote sensing. *Br. phycol. J. 9:*220.

Hughes, J. C., and Lund, J. W. G. 1962. The rate of growth of

Asterionella formosa Hass. in relation to its ecology. *Arch. Mikrobiol.* *42:*117−129.

Hutchens, J. O. 1948. Growth of *Chilomonas paramecium* in mass cultures. *J. cell. comp. Physiol. 32:*105−116.

Hutchinson, G. E. 1944. Limnological studies in Connecticut. VII. A critical examination of the supposed relationship between phytoplankton periodicity and chemical changes in lake waters. *Ecology 25:*3−26.

Hutchinson, G. E. 1957. *A Treatise on Limnology,* vol. 1; New York; 1015 pp.

Hutchinson, G. E. 1967. *A Treatise on Limnology,* vol. 2; New York; 1115 pp.

Hutner, S. H., Baker, H., Aaronson, S., Nathan, H. A., Rodriguez, E., Lockwood, S., Sanders, M., and Peterson, R. A. 1957. Growing *Ochromonas malhamensis* above 35°C. *J. Protozool. 4:*259−269.

Ignatiades, L. 1969. Annual cycle, species diversity and succession of phytoplankton in lower Saronicos Bay, Aegean Sea. *Mar. Biol. 3:*196−200.

Ilmavirta, V., and Hakala, I. 1972. Acrylic plastic and Jena glass bottles used in measuring phytoplanktonic primary production by the ^{14}C method. *Ann. Bot. Fenn. 9:*77−84.

Ingold, C. T. 1971. An olpidioid fungus in the marine diatom *Chaetoceros. Trans. Br. mycol. Soc. 56:*475−486.

Jakob, H. 1961. *Compatibilités, antagonismes et antibioses entre quelques algues du sol.* These no. 4485. Fac. Sciences, Univ. de Paris.

Jensen, A., and Rystad, B. 1973. Semi-continuous monitoring of the capacity of sea water for supporting growth of phytoplankton. *J. exp. mar. Biol. Ecol. 11:*275−285.

Jensen, A., Rystad, B., and Skoglund, L. 1972. The use of dialysis culture in phytoplankton studies. *J. exp. mar. Biol. Ecol. 8:*241−248.

Jerlov, N. G. 1966. Aspects of light measurements in the sea. In *Light as an Ecological Factor,* ed. R. Bainbridge, G. C. Evans, and O. Rackham; Oxford; p. 91−98.

Jerlov, N. G. 1968. *Optical Oceanography.* Amsterdam; 194 pp.

Jitts, H. R., McAllister, C. D., Stephens, K., and Strickland, J. D. H. 1964. The cell division rates of some marine phytoplankters as a function of light and temperature. *J. Fish. Res. Bd Can. 21:*139−157.

Johnson, M. W., and Brinton, E. 1963. Biological species, water-masses and currents. In *The Sea,* vol. 2, ed. M. N. Hill; New York and London; p. 381−414.

Johnston, R. 1963*a.* Antimetabolites as an aid to the study of phytoplankton nutrition. *J. mar. biol. Ass. U.K. 43:*409−425.

Johnston, R. 1963*b*. Sea water, the natural medium of phytoplankton I. General features. *J. mar. biol. Ass. U.K. 43:*427–456.

Jørgensen, E. G. 1956. Growth inhibiting substances formed by algae. *Physiol. Pl. 9:*712–726.

Jørgensen, E. G. 1957. Diatom periodicity and silicon assimilation. *Dansk bot. Ark. 18*(1):1–54.

Jørgensen, E. G. 1962. Antibiotic substances from cells and culture solutions of unicellular algae with special references to some chlorophyll derivatives. *Physiol. Pl. 15:*530–545.

Jørgensen, E. G. 1969. The adaptation of plankton algae IV. Light adaptation in different algal species. *Physiol. Pl. 22:*1307–1315.

Kain, J. M., and Fogg, G. E. 1958. Studies on the growth of marine phytoplankton. I. *Asterionella japonica* Gran. *J. mar. biol. Ass. U.K. 37:*397–413.

Kain, J. M., and Fogg, G. E. 1960. Studies on the growth of marine phytoplankton. III. *Prorocentrum micans* Ehrenberg. *J. mar. biol. Ass. U.K. 39:*33–50.

Kanazawa, T., Kanazawa, K., Kirk, M. R., and Bassham, J. A. 1970. Regulation of photosynthetic carbon metabolism in synchronously growing *Chlorella pyrenoidosa. Pl. Cell Physiol., Tokyo 11*:149–160.

Kessler, E. 1972. Physiologische und biochemische Beiträge zur Taxonomic der Gattung *Chlorella.* VI. Verwertung organischer Kohlenstoff–Verbindungen. *Arch. Mikrobiol. 85*:153–158.

Ketchum, B. H. 1954. Mineral nutrition of phytoplankton. *A. Rev. Pl. Physiol. 5:*55–74.

Ketchum, B. H., Ryther, J. H., Yentsch, C. S., and Corwin, N. 1958. Productivity in relation to nutrients. *Rapp. P.-v. Reun. Cons. perm. int. Explor. Mer 144:*132–140.

Kiefer, D. A. 1973. Fluorescence properties of natural phytoplankton populations. *Mar. Biol. 22:*263–269.

Kratz, W. A., and Myers, J. 1955. Nutrition and growth of several blue-green algae. *Am. J. Bot. 42:*282–287.

Krauss, R. W. 1953. Inorganic nutrition of algae. In *Algal culture from laboratory to pilot plant,* ed. J. S. Burlew; Carnegie Institution of Washington Publication no. 600; p. 85–102.

Kristiansen, J. 1971. Phytoplankton of two Danish lakes with special reference to seasonal cycles of the nannoplankton. *Mitt. int. Verein. theor. angew. Limnol. 19:*253–265.

Kuenzler, E. J., and Ketchum, B. H. 1962. Rate of phosphorus uptake by *Phaeodactylum tricornutum. Biol. Bull. mar. biol. Lab., Woods-Hole 123*: 134–145.

Kumar, H. D. 1964. Streptomycin- and penicillin-induced inhibition of growth and pigment production in blue-green algae and production of strains of *Anacystis nidulans* resistant to these antibiotics. *J. exp. Bot.* *15:*232–250.

Legendre, L., and Watt, W. D. 1970. The distribution of primary production relative to a cyclonic gyre in Baie des Chaleurs. *Mar. Biol.* *7:*167–170.

Lewin, J. C. 1963. Heterotrophy in marine diatoms. In *Symposium on Marine Microbiology,* ed. C. H. Oppenheimer; Springfield, Ill.; p. 229–235.

Lewin, R. A. 1958. Vitamin-bezonoj de algoj. In *Sciencaj Studoj,* ed. P. Neergaard; Copenhagen; p. 187–192.

Lewin, R. A. 1962. (Editor) *Physiology and Biochemistry of Algae.* New York and London.

Lipps, M. J. 1973. The determination of the far-red effect in marine phytoplankton. *J. Phycol. 9:*237–243.

Lund, J. W. G. 1949. Studies on *Asterionella.* I. The origin and nature of the cells producing seasonal maxima. *J. Ecol. 37:*389–419.

Lund, J. W. G. 1950. Studies on *Asterionella formosa* Hass. II. Nutrient depletion and the spring maximum. *J. Ecol. 38:*1–14, 15–35.

Lund, J. W. G. 1954. The seasonal cycle of the plankton diatom, *Melosira italica* (Ehr.) Kütz. subsp. *subarctica* O. Müll. *J. Ecol. 42:*151–179.

Lund, J. W. G. 1955. Further observations on the seasonal cycle of *Melosira italica* (Ehr.) Kütz. subsp. *subarctica* O. Müll. *J. Ecol. 43:*90–102.

Lund, J. W. G. 1959. Buoyancy in relation to the ecology of the freshwater phytoplankton. *Brit. Phycol. Bull. 1*(7):1–17.

Lund, J. W. G. 1961. The periodicity of μ-algae in three English lakes. *Verh. int. Verein. theor. angew. Limnol. 14:*147–154.

Lund, J. W. G. 1971. An artificial alteration of the seasonal cycle of the plankton diatom *Melosira italica* subsp. *subarctica* in an English lake. *J. Ecol. 59:*521–533.

Lund, J. W. G. 1972. Preliminary observations on the use of large experimental tubes in lakes. *Verh. int. Verein. theor. angew. Limnol. 18:*71–77.

Lund, J. W. G., Jaworski, G. H. M., and Bucka, H. 1971. A technique for bioassay of freshwater, with special reference to algal ecology. *Acta hydrobiol. Kraków. 13:*235–249.

Lund, J. W. G., Mackereth, F. J. H., and Mortimer, C. H. 1963. Changes in depth and time of certain chemical and physical conditions and of the standing crop of *Asterionella formosa* Hass. in the north basin of Windermere in 1947. *Phil. Trans. R. Soc., B 246:*255—290.

Lund, J. W. G., and Talling, J. F. 1957. Botanical limnological methods with special reference to the algae. *Bot. Rev. 23:*489—583.

Maaløe, O. 1962. Synchronous growth. In *The Bacteria.* Vol. 4. *The Physiology of Growth,* ed. I. C. Gunsalus and R. Y. Stanier; New York and London; p. 1—32.

Mackereth, F. J. 1953. Phosphorus utilization by *Asterionella formosa* Hass. *J. exp. Bot. 4:*296—313.

Maddux, W. S., and Jones, R. F. 1964. Some interactions of temperature, light intensity, and nutrient concentration during the continuous culture of *Nitzschia closterium* and *Tetraselmis* sp. *Limnol. Oceanogr. 9:*79—86.

Margalef, R. 1958. Temporal succession and spatial heterogeneity in phytoplankton. In *Perspectives in Marine Biology,* ed. A. A. Buzzati-Traverso; Berkeley and Los Angeles; p. 323—349.

Margalef, R. 1968. *Perspectives in Ecological Theory.* Chicago and London; 111 pp.

Mathiesen, H. 1971. Summer maxima of algae and eutrophication. *Mitt. int. Verein. theor. angew. Limnol. 19:*161—181.

McAllister, C. D. 1970. Zooplankton rations, phytoplankton mortality and the estimation of marine production. In *Marine Food Chains,* ed. J. H. Steele; Edinburgh; p. 419—457.

McAllister, C. D., Shah, N., and Strickland, J. D. H. 1964. Marine phytoplankton photosynthesis as a function of light intensity: a comparison of methods. *J. Fish. Res. Bd Can. 21:*159—181.

McLachlan, J. 1960. The culture of *Dunaliella tertiolecta* Butcher—a euryhaline organism. *Can. J. Microbiol. 6:*367—379.

Menzel, D. W., and Spaeth, J. P. 1962. Occurrence of vitamin B_{12} in the Sargasso Sea. *Limnol. Oceanogr. 7:*151—154.

Mihara, S., and Hase, E. 1971. Studies on the vegetative life cycle of *Chlamydomonas reinhardi* Dangeard in synchronous culture I. Some characteristics of the cell cycle. *Pl. Cell Physiol., Tokyo 12:*225—236.

Miller, J. D. H., and Fogg, G. E. 1957. Studies on the growth of Xanthophyceae in pure culture. I. The mineral nutrition of *Monodus subterraneus* Petersen. *Arch. Mikrobiol. 28:*1—17.

Mommearts, J. P. 1973. The relative importance of nannoplankton in the North Sea primary productivity. *Br. phycol. J. 8:*13—20.

Monod, J. 1942. *La croissance des cultures bactériennes.* Paris.

Monod, J. 1950. La technique de culture continue; théorie et applications. *Annls Inst. Pasteur, Paris 79:*390—401.

Morris, I., and Glover, H. E. 1974. Questions on the mechanism of temperature adaptation in marine phytoplankton. *Mar. Biol.* *24:*147–154.

Moss, B. 1969*a*. Vertical heterogeneity in the water column of Abbot's Pond. II. The influence of physical and chemical conditions on the spatial and temporal distribution of the phytoplankton and of a community of epipelic algae. *J. Ecol. 57:*397–414.

Moss, B. 1969*b*. Limitation of algal growth in some central African waters. *Limnol. Oceanogr. 14:*591–601.

Moss, B. 1972. The influence of environmental factors on the distribution of freshwater algae: an experimental study. I. Introduction and the influence of calcium concentration. *J. Ecol. 60:*917–932.

Moss, B. 1973*a*. The influence of environmental factors on the distribution of freshwater algae: an experimental study. II. The role of pH and the carbon dioxide bicarbonate system. *J. Ecol. 61:*157–177.

Moss, B. 1973*b*. The influence of environmental factors on the distribution of freshwater algae: an experimental study. III. Effects of temperature, vitamin requirements and inorganic nitrogen compounds on growth. *J. Ecol. 61:*179–192.

Moss, B. 1973*c*. The influence of environmental factors on the distribution of freshwater algae: an experimental study. IV. Growth of test species in natural waters, and conclusion. *J. Ecol. 61:*193–211.

Moss, B. 1973*d*. Diversity in fresh-water phytoplankton. *Am. Midl. Nat. 90:*341–355.

Munk, W. H., and Riley, G. A. 1952. Absorption of nutrients by aquatic plants. *J. mar. Res. 11:*215–240.

Myers, J. 1946*a*. Culture conditions and the development of the photosynthetic mechanism. III. Influence of light intensity on cellular characteristics of *Chlorella. J. gen. Physiol. 29:*419–427.

Myers, J. 1946*b*. Culture conditions and the development of the photosynthetic mechanism. IV. Influence of light intensity on photosynthetic characteristics of *Chlorella. J. gen. Physiol. 29:*429–440.

Myers, J. 1947. Culture conditions and the development of the photosynthetic mechanism. V. Influence of the composition of the nutrient medium. *Pl. Physiol. 22:*590–597.

Myers, J. 1951. Physiology of the Algae. *A. Rev. Microbiol. 5:*157–180.

Myers, J. 1953. Growth characteristics of algae in relation to the problems of mass culture. In *Algal Culture from Laboratory to Pilot Plant,* ed. J. S. Burlew; Carnegie Institution of Washington Publication no. 600; p. 37–54.

Myers, J. 1962*a*. Laboratory Cultures. In *Physiology and Biochemistry of Algae*, ed. R. A. Lewin; New York and London; p. 603–615.

Myers, J. 1962*b*. Variability of metabolism in algae. In *Beiträge zur Physiologie und Morphologie der Algen.* Vorträge aus dem Gesamtgebiet der Botanik, Deutschen Botanischen Gesellschaft N.F. *1:*13–19.

Myers, J., and Clark, L. B. 1944. Culture conditions and the development of the photosynthetic mechanism. II. An apparatus for the continuous culture of *Chlorella. J. gen. Physiol. 28:*103–112.

Myers, J., and Graham, J.-R. 1956. The role of photosynthesis in the physiology of *Ochromonas. J. cell. comp. Physiol. 47:*397–414.

Myers, J., and Graham, J.-R. 1959. On the mass culture of algae. II. Yield as a function of cell concentration under continous sunlight irradiance. *Pl. Physiol. 34:*345–352.

Myers, J., Phillips, J. N., and Graham. J.-R. 1951. On the mass culture of algae. *Pl. Physiol. 26:*539–548.

Nakamura, K., and Gowans, C. S. 1964. Nicotinic acid-excreting mutants in *Chlamydomonas. Nature, Lond. 202:*826–827.

Nalewajko, C., Chowdhuri, N., and Fogg, G. E. 1963. Excretion of glycollic acid and the growth of a planktonic *Chlorella.* In *Studies on Microalgae and Photosynthetic Bacteria.* Tokyo; p. 171–183.

Nauwerck, A. 1963. Die Beziehungen zwischen Zooplankton und Phytoplankton im See Erken. *Symb. bot. upsal. 17*(5):1–163.

Nihei, T., Sasa, T., Miyachi, S., Suzuki, K., and Tamiya, H. 1954. Change of photosynthetic activity of *Chlorella* cells during the course of their normal life cycle. *Arch. Mikrobiol. 21:*155–164.

Nordli, E. 1957. Experimental studies on the ecology of *Ceratia. Acta oecol. scand. 8:*200–265.

Novick, A., and Szilard, L. 1950. Description of the chemostat. *Science, Washington, D.C. 112:*715–716.

Ohle, W. 1958. Diurnal production and destruction rates of phytoplankton in lakes. *Rapp. P.-v. Reun. Cons. perm. int. Explor. Mer 144:*129–131.

Ohle, W. 1961. Tagesrhythmen der Photosynthese von Planktonbiocoenosen. *Verh. int. Verein. theor. angew. Limnol. 14:*113–119.

O'Kelley, J. C. 1968. Mineral nutrition of algae. *A. Rev. Pl. Physiol. 19:*89–112.

van Oorschot, J. L. P. 1955. Conversion of light energy in algal cultures. *Meded. LandbHoogesch., Wageningen 55:*225–276.

Oppenheimer, C. H. 1966. *Marine Biology II,* ed. C. H. Oppenheimer; New York; 369 pp.

Österlind, S. 1949. Growth conditions of the alga *Scenedesmus quad-ricauda* with special reference to the inorganic carbon sources. *Symb. bot. upsal. 10*(3):1—141.

Otsuka, H. 1961. Changes of lipid and carbohydrate contents in *Chlorella* cells during the sulfur starvation, as studied by the technique of synchronous culture. *J. gen. appl. Microbiol., Tokyo 7*:72—77.

Paasche, E. 1960*a*. On the relationship between primary production and standing stock of phytoplankton. *J. Cons. perm. int. Explor. Mer 26*:33—48.

Paasche, E. 1960*b*. Phytoplankton distribution in the Norwegian Sea in June, 1954, related to hydrography and compared with primary production data. *Fiskeridir. Skr. Havundersøk. 12*(11): 1—77.

Paasche, E. 1964. A tracer study of the inorganic carbon uptake during coccolith formation and photosynthesis in the coccolithophorid *Coccolithus huxleyi. Physiol. Pl. Suppl. 3*:5—82.

Paasche, E. 1967. Marine plankton algae grown with light-dark cycles. I. *Coccolithus huxleyi. Physiol. Pl. 20*:946—956.

Paasche, E. 1968. Marine plankton algae grown with light-dark cycles. II. *Ditylum brightwellii* and *Nitzschia turgidula. Physiol. Pl. 21*:66—77.

Pant, A. 1973. Uptake of an extracellular product, glycollic acid, by a neritic photosynthesizing species, *Skeletonema costatum* (Grev.) Cleve, in culture and in the sea. Unpublished Ph.D. diss., University of London.

Paredes, J. F. 1962. On an occurrence of red waters in the coast of Angola. *Mems Jta Invest. Ultramar* 2nd. ser., no. 33, p. 89—114.

Paredes, J. F. 1967/68*a*. Studies on cultures of marine phytoplankton. I. Diatom *Ditylum brightwellii* West. *Mems Inst. Invest. cient. Moçamb.* Série A, *9*:157—183.

Paredes, J. F. 1967/68*b*. Studies on cultures of marine phytoplankton. III. Chlorophycean *Micromonas squamata* Manton & Parke. *Mems Inst. Invest. cient. Moçamb.*, Série A, *9*:249—292.

Parker, B. C., Leeper, G., and Hurni, W. 1968. Sampler for studies of thin horizontal layers. *Limnol. Oceangr. 13*:172—175.

Parker, B. C., and Wachtel, M. A. 1971. Seasonal distribution of cobalamins, biotin, and niacin in rainwater. In *The structure and function of fresh-water microbial communities,* ed. J. Cairns, Jr.; Blacksburg, Va.; Virginia Polytechnic Institute and State University Research Division Monograph 3; p. 195—207.

Parsons, T. R., Stephens, K., and Strickland, J. D. H. 1961. On the chemical composition of eleven species of marine phytoplankters. *J. Fish. Res. Bd Can. 18:*1001–1016.

Parsons, T. R., and Strickland, J. D. H. 1962. On the production of particulate organic carbon by heterotrophic processes in sea water. *Deep Sea Res. 8:*211–222.

Pearsall, W. H. 1932. Phytoplankton in the English Lakes. 2. The composition of the phytoplankton in relation to dissolved substances. *J. Ecol. 20:*241–262.

Pearsall, W. H., and Bengry, R. P. 1940. The growth of *Chlorella* in darkness and in glucose solution. *Ann. Bot. N.S. 4:*365–377.

Pechlaner, R. 1970. The phytoplankton spring outburst and its conditions in Lake Erken (Sweden). *Limnol. Oceanogr. 15:*113–130.

Phillips, J. N., Jr., and Myers, J. 1954. Measurement of algal growth under controlled steady-state conditions. *Pl. Physiol. 29:*148–152.

Pintner, I. J., and Provasoli, L. 1963. Nutritional characteristics of some chrysomonads. In *Symposium on Marine Microbiology,* ed. C. H. Oppenheimer; Springfield, Ill.; p. 114–121.

Pirson, A. 1957. Induced periodicity of photosynthesis and respiration in *Hydrodictyon.* In *Research in Photosynthesis,* ed. H. Gaffron *et al.;* New York and London; p. 490–499.

Pirson, A., and Lorenzen, H. 1958. Ein endogener Zeitfaktor bei der Teilung von *Chlorella. Z. Bot. 46:*53–66.

Pirson, A. and Lorenzen, H. 1966. Synchronized dividing algae. *A. Rev. Pl. Physiol. 17:*439–458.

Platt, T., and Subba Rao, D. V. 1970. Primary production measurements on a natural plankton bloom. *J. Fish. Res. Bd Can. 27:*887–899.

Platt, T., and Subba Rao, D. V. 1973. Some current problems in marine phytoplankton productivity. Fisheries Research Board of Canada, Technical Report no. 370. To be published in *The Functioning of Photosynthetic Systems in Different Environments,* ed. J. P. Cooper; New York.

Pratt, D. M. 1966. Competition between *Skeletonema costatum* and *Olisthodiscus luteus* in Narragansett Bay and in culture. *Limnol. Oceanogr. 11:*447–455.

Pratt, R. 1943. Studies on *Chlorella vulgaris.* VI. Retardation of photosynthesis by a growth-inhibiting substance from *Chlorella vulgaris. Am. J. Bot. 30:*32–33.

Pratt, R., and Fong, J. 1940. Studies on *Chlorella vulgaris.* II. Further evidence that *Chlorella* cells form a growth-inhibiting substance. *Am. J. Bot. 27:*431–436.

Pringsheim, E. G. 1946. *Pure Cultures of Algae*. Cambridge.

Pringsheim, E. G., and Wiessner, W. 1961. Ernährung und Stoffwechsel von *Chlamydobotrys* (Volvocales). *Arch. Mikrobiol. 40:*231–246.

Proctor, V. W. 1957. Studies of algal antibiosis using *Haematococcus* and *Chlamydomonas. Limnol. Oceanogr. 2:*125–139.

Provasoli, L. 1963. Organic regulation of phytoplankton fertility. In *The Sea*, vol. 2, ed. M. N. Hill; New York and London; p. 165–219.

Provasoli, L. 1966. Media and prospects for the cultivation of marine algae. In *Cultures and Collections of Algae*, ed. A. Watanabe & A. Hattori; Tokyo; p. 63–75.

Provasoli, L. 1971. Nutritional relationships in marine organisms. In *Fertility of the Sea*, ed. J. D. Costlow, Jr.; New York; p. 369–382.

Provasoli, L., McLaughlin, J. J. A., and Droop, M. R. 1957. The development of artificial media for marine algae. *Arch. Mikrobiol. 25:*392–428.

Qasim, S. Z., Wellershaus, S., Bhattathiri, P. M. A., and Abidi, S. A. H. 1969. Organic production in a tropical estuary. *Proc. Indian. Acad. Sci. B. 69:*51–94.

Quraishi, F. O., and Spencer, C. P. 1971. Studies on the responses of marine phytoplankton to light fields of varying intensity. In *Fourth European Marine Biology Symposium*, ed. D. J. Crisp; New York; p. 393–408.

Raven, J. A. 1970. Exogenous inorganic carbon sources in plant photosynthesis. *Biol. Rev. 45:*167–221.

Raymont, J. E. G. 1963. *Plankton and Productivity in the Oceans*. London; 660 pp.

Reid, F. M. H., Fuglister, E., and Jordan, J. B. 1970. The ecology of the plankton off La Jolla California, in the period April through September, 1967. Part V. Phytoplankton taxonomy and standing crop. *Bull. Scripps Instn Oceanogr. non-tech. Ser. 17:*51–66.

Reynolds, C. S. 1971. The ecology of the planktonic blue-green algae in the North Shropshire meres. *Fd. Stud. 3:*409–432.

Reynolds, C. S. 1973. Growth and buoyancy of *Microcystis aeruginosa* Kütz. emend. Elenkin in a shallow eutrophic lake. *Proc. R. Soc., B 184:*29–50.

Reynolds, N. 1973. The estimation of the abundance of ultraplankton. *Br. phycol. J. 8:*135, 146.

Riley, G. A. 1963. Theory of food-chain relations in the ocean. In *The Sea*, vol. 2, ed. M. N. Hill; New York and London; p. 438–463.

Roach, B. M. Bristol. 1928. On the influence of light and of glucose on the growth of a soil alga. *Ann. Bot. 42:*317–345.

Rodhe, W. 1948. Environmental requirements of fresh-water plankton algae. *Symb. bot. upsal. 101:*1–149.

Rodhe, W. 1955. Can plankton production proceed during winter darkness in sub-arctic lakes? *Verh. int. Verein. theor. angew. Limnol. 12:*117–122.

Rodhe, W. 1958. The primary production in lakes: some results and restrictions of the ^{14}C method. *Rapp. P.-v. Reun. Cons. Perm. int. Explor. Mer 144*:122–128.

Rodhe, W., Vollenweider, R. A., and Nauwerck, A. 1958. The primary production and standing crop of phytoplankton. In *Perspectives in Marine Biology,* ed. A. A. Buzzati-Traverso; Berkeley and Los Angeles; p. 299–322.

Round, F. E. 1971. The growth and succession of algal populations in freshwaters. *Mitt. int. Verein. theor. angew. Limnol. 19:*70–99.

Ruttner, F. 1953. *Fundamentals of Limnology.* Toronto.

Ryther, J. H. 1955. Ecology of autotrophic marine dinoflagellates with reference to red water conditions. In *The Luminescence of Biological Systems,* ed. F. H. Johnson; Washington, D.C.; p. 387–414.

Ryther, J. H. 1963. Geographic variations in productivity. In *The Sea,* vol. 2, ed. M. N. Hill; New York and London; p. 347–380.

Ryther, J. H., and Guillard, R. R. L. 1959. Enrichment experiments as a means of studying nutrients limiting to phytoplankton production. *Deep Sea Res. 6:*65–69.

Ryther, J. H., and Menzel, D. W. 1959. Light adaptation by marine phytoplankton. *Limnol. Oceanogr. 4:*492–497.

Schindler, D. W. 1971. Carbon, nitrogen, and phosphorus and the eutrophication of freshwater lakes. *J. Phycol. 7:*321–329.

Sen, N., and Fogg, G. E. 1966. Effects of glycollate on the growth of a planktonic *Chlorella. J. exp. Bot. 17:*417–425.

Senger, H. 1970a. Charakterisierung einer Synchronkultur von *Scenedesmus obliquus* ihrer potentiellen Photosyntheseleistung und des Photosynthese—Quotienten während des Entwicklungscycles. *Planta (Berl.) 90:*243–266.

Senger, H. 1970b. Quantenausbeute und unterschiedliches Verhalten der beiden Photosysteme des Photosyntheseapparates während des Entwicklungsablaufes von *Scenedesmus obliquus* in Synchronkulturen. *Planta (Berl.) 92:*327–346.

Shah, N. M. 1973. Seasonal variation of phytoplankton pigments and some of the associated oceanographic parameters in the Laccadive Sea off Cochin. In *The Biology of the Indian Ocean,* ed. B. Zeitzschel and S. A. Gerlach; London; p. 175–185.

Shapiro, J. 1973. Carbon dioxide and pH: effect on species succession of algae. *Science 182:*307.

Shihira, I., and Krauss, R. W. *circa* 1965. *Chlorella. Physiology and Taxonomy of Forty-one Isolates.* College Park, Md.; 97 pp.

Shimada, B. M. 1958. Diurnal fluctuation in photosynthetic rate and chlorophyll A content of phytoplankton from eastern Pacific waters. *Limnol. Oceanogr. 3:*336–339.

Sieburth, J. M. 1968. The influence of algal antibiosis on the ecology of marine microorganisms. In *Advances in Microbiology of the Sea,* vol. 1, ed. M. R. Droop and E. J. Ferguson Wood; London and New York; p. 63–94.

Smayda, T. J. 1958. Biogeographical studies of marine phytoplankton. *Oikos 9:*158–191.

Smayda, T. J. 1963. Succession of phytoplankton, and the ocean as an holocoenotic environment. In *Symposium on Marine Microbiology,* ed. C. H. Oppenheimer; Springfield, Ill.; p. 260–274.

Smayda, T. J. 1970. The suspension and sinking of phytoplankton in the sea. *Oceanogr. & mar. Biol. Rev. 8:*353–414.

Smayda, T. J. 1973. The growth of *Skeletonema costatum* during a winter-spring bloom in Narragansett Bay, Rhode Island. *Norw. J. Bot. 20:*219–247.

Soeder, C. J. 1970. Zum Phosphat-Haushalt von *Chlorella fusca* Sh. et. Kr. *Arch. Hydrobiol., Suppl. 38:*1–17.

Soeder, C. J., Müller, H., Payer, H. D., and Schulle, H. 1971. Mineral nutrition of planktonic algae: some considerations, some experiments. *Mitt int. Verein. theor. angew. Limnol. 19:*39–58.

Soeder, C. J., and Ried, A. 1967. Über die Atmung synchron kultivierter *Chlorella.* II. Die Entwicklungsabhängigkeit der Veränderungen des respiratorischen Gaswechsels. *Arch. Mikrobiol. 56:*106–119.

Sorokin, C. 1959. Tabular comparative data for the low- and high-temperature strains of *Chlorella. Nature, Lond. 184:*613–614.

Sorokin, C. 1964. Organic synthesis in algal cells separated into age groups by fractional contrifugation. *Arch. Mikrobiol. 49:*193–208.

Sorokin, C., and Myers, J. 1953. A high temperature strain of *Chorella. Science, Washington, D.C. 117:*330–331.

Sournia, A. 1967. Rythme nycthéméral du rapport "intensite photo-synthetique/chlorophylle" dans le plancton marin. *C. r. hebd. Séanc. Acad. Sci., Paris 265:*1000–1003.

Spektorov, K. S., and Lin'kova, E. A. 1962. A new simple method of synchronizing *Chlorella cultures. Dokl. Akad. Nauk SSSR 147:*967–969.

Spencer, C. P. 1954. Studies on the culture of a marine diatom. *J. mar. biol. Ass. U.K. 33:*265–290.

Spoehr, H. A., and Milner, H. W. 1949. The chemical composition of *Chlorella;* effect of environmental conditions. *Pl. Physiol. 24:*120–149.

Staub, R. 1961. Ernährungsphysiologisch-autökologische Untersuchungen an der planktischen Blaualge *Oscillatoria rubescens* DC. *Schweiz. Z. Hydrol. 23:*82–198a.

Steele, J. H. 1958. *Plant production in the northern North Sea.* Scottish Home Department, Marine Research no. 7; Edinburgh.

Steele, J. H. 1962. Environmental control of photosynthesis in the sea. *Limnol. Oceanogr. 7:*137–150.

Steele, J. H. 1963. In *Marine Biology I. First International Interdisciplinary Conference on Marine Biology,* ed. G. A. Riley; Washington, D.C.; p. 50.

Steele, J. H., and Baird, I. E. 1961. Relations between primary production, chlorophyll and particulate carbon. *Limnol. Oceanogr. 6:*68–78.

Steele, J. H., and Baird, I. E. 1962. Carbon-chlorophyll relations in cultures. *Limnol. Oceanogr. 7:*101–102.

Steele, J. H., and Yentsch, C. S. 1960. The vertical distribution of chlorophyll. *J. mar. biol. Ass. U.K. 39:*217–226.

Steemann Nielsen, E. 1955. The production of organic matter by the phytoplankton in a Danish lake receiving extraordinarily great amounts of nutrient salts. *Hydrobiologia 7:*68–74.

Steeman Nielsen, E. 1963. Productivity, definition and measurement. In *The Sea,* vol. 2, ed. M. N. Hill; New York and London; 129–164.

Steeman Nielsen, E. 1964a. Recent advances in measuring and understanding marine primary production. *J. Ecol. 52* (Suppl.):119–130.

Steeman Nielsen, E. 1964b. Investigations of the rate of primary production at two Danish light ships in the transition area between the North Sea and the Baltic. *Meddr Danm. Fisk. og Havunders.,* N.S., 4:31–77.

Steeman Nielsen, E., and Hansen, V. K. 1959. Light adaptation in marine phytoplankton populations and its interrelation with temperature. *Physiol. Pl. 12:*353–370.

Steeman Nielsen, E., Hansen, V. K., and Jørgensen, E. G. 1962. The adaptation to different light intensities in *Chlorella vulgaris* and the time dependence on transfer to a new light intensity. *Physiol. Pl. 15:*505–517.

Steeman Nielsen, E., and Jørgensen, E. G. 1962. The physiological background for using chlorophyll measurements in hydrobiology and a theory explaining daily variations in chlorophyll concentration. *Arch. Hydrobiol. 58:*349–357.

Steeman Nielsen, E., and Jørgensen, E. G. 1968. The adaptation of algae. I. General Part. *Physiol. Pl. 21*:401–413.

Stein, J. R. 1973. *Handbook of Phycological Methods. Culture Methods and Growth Measurements,* ed. J. R. Stein; New York; 448 pp.

Stewart, W. D. P. 1971. Nitrogen fixation in the sea. In *Fertility of the Sea,* vol. 2, ed. J. D. Costlow, Jr.; New York and London; p. 537–564.

Stewart, W. D. P. 1973. Nitrogen fixation. In *The Biology of Blue-green Algae,* ed. N. G. Carr and B. A. Whitton; Oxford; p. 260–278.

Stewart, W. D. P. 1974. *Algal Physiology and Biochemistry,* ed. W. D. P. Stewart; Oxford; 960 pp.

Stewart, W. D. P., and Alexander, G. 1971. Phosphorus availability and nitrogenase activity in aquatic blue-green algae. *Freshwat. Biol. 1:*389–404.

Stewart, W. D. P., Fitzgerald, G. P., and Burris, R. H. 1970. Acetylene reduction assay for determination of phosphorus availability in Wisconsin lakes. *Proc. natn. Acad. Sci. U.S.A. 66:*1104–1111.

Stewart, W. D. P., and Pearson, H. W. 1970. Effects of aerobic and anaerobic conditions on growth and metabolism of blue-green algae. *Proc. R. Soc., B 175:*293–311.

Strickland, J. D. H. 1965. In *Chemical Oceanography,* vol. 1, ed. J. P. Riley and G. Skirrow; London and New York; p. 477–610.

Strickland, J. D. H. 1972. Research on the marine planktonic food web at the Institute of Marine Resources: a review of the past seven years of work. *Oceanogr. & mar. Biol. a. Rev. 10:*349–414.

Strickland, J. D. H., Eppley, R. W., and Mendiola, B. R. de. 1969*a*. Phytoplankton populations and photosynthesis in Peruvian coastal waters. *Bol. Inst. Mar. Peru-Callas 2:*4–45.

Strickland, J. D. H., Holm-Hansen, O., Eppley, R. W., and Linn, R. J. 1969*b*. The use of a deep tank in plankton ecology. I. Studies of the growth and composition of phytoplankton crops at low nutrient levels. *Limnol. Oceanogr. 14:*23–34.

Strickland, J. D. H., and Parsons, T. R. 1968. A practical handbook of water analysis. *Fisheries Research Board of Canada Bulletin 167*:311 pp.

Strickland, J. D. H., and Terhune, L. D. B. 1961. The study of *in situ* marine photosynthesis using a large plastic bag. *Limnol. Oceanogr.* 6:93–96.

Stross, R. G., Neess, J. C., and Hasler, A. D. 1961. Turnover time and production of the planktonic crustacea in limed and reference portion of a bog lake. *Ecology 42:*237–245.

Sutcliffe, W. H., Jr., Baylor, E. R., and Menzel, D. W. 1963. Sea surface chemistry and Langmuir circulation. *Deep Sea Res. 10:*233–243.

Sverdrup, H. U. 1953. On conditions for the vernal blooming of phytoplankton. *J. Cons. perm. int. Explor. Mer 18:*287–295.

Sverdrup, H. U., Johnson, M. W., and Fleming, R. H. 1942. *The Oceans.* Englewood Cliffs. N.J.

Sweeney, B. M., and Hastings, J. W. 1962. Rhythms. In *Physiology and Biochemistry of Algae,* ed. R. A. Lewin; New York and London; p. 687–700.

Taguchi, S. 1970. Seasonal variations of photosynthetic behaviour of phytoplankton in Akkeshi Bay, Hokkaido, with special reference to low photosynthetic rate in summer associated with large percentage of dwarf cells. *Bull. Plankton. Soc. Japan 17:*65–77.

Talling, J. F. 1957. The growth of two plankton diatoms in mixed cultures. *Physiol. Pl. 10:*215–223.

Talling, J. F. 1960*a.* Comparative laboratory and field studies of photosynthesis by a marine planktonic diatom. *Limnol. Oceanogr. 5:*62–77.

Talling, J. F. 1960*b.* Self-shading effects in natural populations of a planktonic diatom. *Wett. Leben 12:*235–242.

Talling, J. F. 1961. Report on limnological work during a visit to EAFFRO between August 1960, and September 1961. *East African Freshwater Fishery Research Organization, Annual Report 1961,* p. 40–42.

Talling, J. F. 1962. Freshwater algae. In *Physiology and Biochemistry of Algae,* ed. R. A. Lewin; New York and London; p. 743–757.

Talling, J. F., Wood, R. B., Prosser, M. V., and Baxter, R. M. 1973. The upper limit of photosynthetic productivity by phytoplankton: evidence from Ethiopian soda lakes. *Freshwat. Biol. 3:*53–76.

Tamiya, H. 1963. Cell differentiation in *Chlorella. Symp. Soc. exp. Biol. 17:*188–214.

Tamiya, H. 1964. Growth and cell division of *Chlorella.* In *Synchrony in Cell Division and Growth,* ed. E. Zeuthen; New York; p. 247–305.

Tamiya, H. 1966. Synchronous cultures of algae. *A. Rev. Pl. Physiol. 17:*1–26.

Tamiya, H., Iwamura, T., Shibata, K., Hase, E., and Nihei, T. 1953. Correlation between photosynthesis and light-independent metabolism in the growth of *Chlorella. Biochim. biophys. Acta 12:*23–40.

Tamiya, H., Sasa, T., Nihei, T., and Ishibashi, S. 1955. Effect of variation of day-length, day and night-temperatures, and intensity of daylight upon the growth of *Chlorella. J. gen. appl. Microbiol., Tokyo 1:*298–307.

Tassigny, M., and Lefèvre, M. 1971. Auto., héteroantagonisme et autres conséquences des excrétions d'algues d'eau douce ou thermale. *Mitt. int. Verein. theor. angew. Limnol. 19:*26–38.

Thomas, W. H. 1966. Effects of temperature and illuminance on cell division rates of three species of tropical oceanic phytoplankton. *J. Phycol. 2:*17–22.

Thomas, W. H. 1969. Phytoplankton nutrient enrichment experiments off Baja California and in the eastern equatorial Pacific Ocean. *J. Fish. Res. Bd Can. 26:*1133–1145.

Thomas, W. H., and Dodson, A. N. 1968. Effects of phosphate concentration on cell division rates and yield of a tropical oceanic diatom. *Biol. Bull. mar. biol. Lab., Woods Hole 134:*199–208.

Thomas, W. H., and Dodson, A. N. 1972. On nitrogen deficiency in tropical Pacific oceanic phytoplankton. II. Photosynthetic and cellular characteristics of a chemostat-grown diatom *Limnol. Oceanogr. 17:*515–523.

Tokuda, H. 1966. On the culture of a marine diatom *Nitzschia closterium.* In *Cultures and Collections of Algae,* ed. A. Watanabe and A. Hattori; Tokyo; p. 53–58.

Tolbert, N. E. 1974. Photorespiration. In *Algal Physiology and Biochemistry,* ed. W. D. P. Stewart; Oxford; p. 474–504.

Tolbert, N. E., and Zill, L. P. 1956. Excretion of glycolic acid by algae during photosynthesis. *J. biol. Chem. 222:*895–906.

Van Baalen, C. 1962. Studies on marine blue-green algae. *Botanica mar. 4:*129–139.

Vance, B. D. 1965. Composition and succession of cyanophycean water blooms. *J. Phycol. 1:*81–86.

Verduin, J. 1951. A comparison of phytoplankton data obtained by a mobile sampling method with those obtained from a single station. *Am. J. Bot. 38:*5–11.

Verduin, J. 1956. Energy fixation and utilization by natural communities in western Lake Erie. *Ecology 37:*40–49.

Viner, A. B. 1973. Responses of a mixed phytoplankton population to nutrient enrichments of ammonia and phosphate, and some associated ecological implications. *Proc. R. Soc. Lond., B 183:*351–370.

Vollenweider, R. A. 1968. *Scientific fundamentals of the eutrophication of lakes and flowing waters with particular reference to nitrogen and phosphorus as factors in eutrophication.* Paris.

Vollenweider, R. A. 1974. *A Manual on Methods for Measuring Primary Production in Aquatic Environments,* ed. R. A. Vollenweider; I. B. P. Handbook no. 12; 2d ed.; Oxford; 225 pp.

Walsby, A. E. 1967. A new culture flask. *Biotechnol. & Bioeng 9:*443–447.

Walsby, A. E. 1971. The pressure relationships of gas vacuoles. *Proc. R. Soc. Lond., B 178:*301–326.

Walsby, A. E. 1972. Structure and function of gas-vacuoles. *Bact. Rev. 36:*1–32.

Walsby, A. E. 1973. Gas vacuoles. In *The Biology of Blue-green Algae,* ed. N. G. Carr and B. A. Whitton; Oxford; p. 340–352.

Walsby, A. E., and Klemer, A. R. 1974. The role of gas vacuoles in the microstratification of a population of *Oscillatoria agardhii* var. *isothrix* in Deming Lake, Minnesota. *Arch. Hydrobiol. 74:*375–392.

Wassink, E. C. 1954. Problems in the mass cultivation of photoautotrophic micro-organisms. *Symp. Soc. gen. Microbiol. 4:*247–270.

Watt, W. D., and Hayes, F. R. 1963. Tracer study of the phosphorus cycle in sea water. *Limnol. Oceanogr. 8:*276–285.

Whitton, B. A. 1967. Studies on the toxicity of polymyxin B to blue-green algae. *Can. J. Microbiol. 13:*987–993.

Wiessner, W. 1970. Photométabolism of organic substances. In *Photobiology of Microorganisms,* ed. P. Halldal; London and New York; p. 95–133.

Wilson, D. P., and Armstrong, F. A. J. 1958. Biological differences between sea waters: experiments in 1954 and 1955. *J. mar. biol. Ass. U.K. 37:*331–348.

Winokur, M. 1948. Growth relationships of *Chlorella* species. *Am. J. Bot. 35:*118–129.

Woodcock, A. H. 1950. Subsurface pelagic *Sargassum. J. mar. Res. 9:*77–92.

Wright, J. C. 1960. The limnology of Canyon Ferry Reservoir: III. Some observations on the density dependence of photosynthesis and its cause. *Limnol. Oceanogr. 5:*356–361.

Wright, R. T. 1964. Dynamics of a plytoplankton community in an ice-covered lake. *Limnol. Oceanogr. 9:*163–178.

Yentsch, C. S., and Ryther, J. H. 1957. Short-term variations in phytoplankton chlorophyll and their significance. *Limnol. Oceanogr.* *2:*140–142.

Zavarzina, N. B. 1959. On some factors inhibiting the development of *Scenedesmus quadricauda. Trudȳ vses. gidrobiol. Obshch. 9:*195–205.

Zehnder, A. 1963. Kulturversuche mit *Gloeotrichia echinulata* (J. E. Smith) P. Richter. *Schweiz. Z. Hydrol. 25:*65–83.

Zimmermann, U. 1969. Ökologische und physiologische Untersuchungen an der planktonischen Blaualge *Oscillatoria rubescens.* D.C., unter besonderer Berücksichtigung von Licht und Temperatur. *Schweiz. Z. Hydrol. 31:*1–58.

ZoBell, C. E. 1946. *Marine Microbiology.* Waltham, Mass.

Index

167

170 Index

DESIGNED BY IRVING PERKINS
COMPOSED BY THE COMPOSING ROOM, GRAND RAPIDS, MICHIGAN
MANUFACTURED BY CUSHING-MALLOY, INC., ANN ARBOR, MICHIGAN
TEXT IS SET IN PRESS ROMAN, DISPLAY LINES IN ALTERNATE GOTHIC

Library of Congress Cataloging in Publication Data

Fogg, Gordon Elliott.
Algal cultures and phytoplankton ecology.

Bibliography: p. 141-166.
Includes index.
1. Algae—Cultures and culture media. 2. Phyto-
plankton—Ecology. I. Title.
QK565.2.F63 1975   589'.3   74-27308
ISBN 0-299-06760-2